PRACTICAL PROBLEMS in MATHEMATICS
for GRAPHIC COMMUNICATIONS
Second Edition

Delmar's *PRACTICAL PROBLEMS in MATHEMATICS* Series

- *Practical Problems in Mathematics for Automotive Technicians, 5e*
George Moore
Revised by Larry and Todd Sformo
Order # 0-8273-7944-7

- *Practical Problems in Mathematics for Carpenters, 6e*
Harry C. Huth
Order # 0-8273-4579-8

- *Practical Problems in Mathematics for Drafting and CAD, 2e*
John C. Larkin
Order # 0-8273-1670-4

- *Practical Problems in Mathematics for Electricians, 5e*
Herman and Garrard
Order # 0-8273-6708-2

- *Practical Problems in Mathematics for Electronics Technicians, 5e*
Herman and Sullivan
Order # 0-8273-6761-9

- *Practical Problems in Mathematics for Graphic Communications, 2e*
Ervin A. Dennis
Order # 0-8273-7946-3

- *Practical Problems in Mathematics for Health Occupations*
Louise M. Simmers
Order # 0-8273-6771-6

- *Practical Problems in Mathematics for Heating and Cooling Technicians, 3e*
Russell B. DeVore
Order # 0-8273-7948-X

- *Practical Problems in Mathematics for Industrial Technology*
Donna Boatwright
Order # 0-8273-6974-3

- *Practical Problems in Mathematics for Manufacturing, 4e*
Dennis D. Davis
Order # 0-8273-6710-4

- *Practical Problems in Mathematics for Masons, 2e*
John E. Ball
Order # 0-8273-1283-0

- *Practical Problems in Mathematics for Welders, 4e*
Schell and Matlock
Order # 0-8273-6706-6

Online Services

Delmar Online
To access a wide variety of Delmar products and services on the World Wide Web, point your browser to:
> http://www.delmar.com/delmar.html
> or email: info@delmar.com

thomson.com
To access International Thomson Publishing's home site for information on more than 34 publishers and 20,000 products, point your browser to:
> http://www.thomson.com
> or email: findit@kiosk.thomson.com

A service of I(T)P®

PRACTICAL PROBLEMS in MATHEMATICS
for GRAPHIC COMMUNICATIONS
Second Edition

Ervin A. Dennis, Ed.D.
LaVonne Vermeersch, and
Charles Southwick

Delmar Publishers Inc.

I**T**P An International Thomson Publishing Company

Albany • Bonn • Boston • Cincinnati • Detroit • London • Madrid • Melbourne
Mexico City • New York • Pacific Grove • Paris • San Francisco • Singapore • Tokyo
Toronto • Washington

NOTICE TO THE READER

Cover Design: John Kenific

Delmar Staff
Publisher: Alar Elken
Administrative Editor: Paul Shepardson
Developmental Editor: Julie Waite
Production Manager: Larry Main
Art and Design Coordinator: Nicole Reamer

COPYRIGHT © 1998
by Delmar Publishers
A division of International Thomson Publishing Inc.
The ITP logo is a trademark under license.

Printed in the United States of America

For more information contact:
Delmar Publishers
3 Columbia Circle, Box 15015
Albany, New York 12212-5015

International Thomson Editores
Campos Eliseos 385, Piso 7
Col Polanco
11560 Mexico D F Mexico

International Thomson Publishing - Europe
Berkshire House 168-173
High Holborn
London, WCIV 7AA
England

International Thomson Publishing GmbH
Königswinterer Strasse 418
53227 Bonn
Germany

Thomas Nelson Australia
102 Dodds Street
South Melbourne, 3205
Victoria, Australia

International Thomson Publishing - Asia
221 Henderson Road
#05-10 Henderson Building
Singapore 0315

Nelson Canada International
1120 Birchmount Road
Scarborough, Ontario
Canada, M1K 5G4

International Thomson Publishing - Japan
Hirakawacho Kyowa Building, 3F
2-2-1 Hirakawacho
Chiyoda-ku, Tokyo 102
Japan

4 5 6 7 8 9 10   xxx   04 03 02

Library of Congress Cataloging-in-Publication Data
Dennis, Ervin A.
    Practical problems in mathematics for graphic communications /
Ervin A. Dennis, LaVonne Vermeersch, and Charles Southwick — 2nd ed.
        p.   cm. — (Delmar's practical problems in mathematics series)
        ISBN 0-8273-7946-3 (alk. paper)
        1. Printing—Mathematics—Problems, exercises, etc.
    2. Graphic arts—Mathematics—Problems, exercises, etc.
    I. Vermeersch, LaVonne.   II. Southwick, Charles.   III. Title.
Z253.D46   1997   97-28541
686.2'01'51—dc20   CIP

Contents

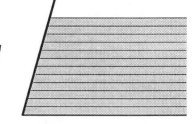

Preface

Mathematics is an important academic discipline to know for anyone who wishes to be successful in the technical and professional area of graphic communications. All areas of graphic communications require basic knowledge in the practices and procedures of mathematics. Designers, computer operators, press operators, and finishing and binding personnel must know how to work with whole numbers, common fractions, decimal fractions, percent, and measurement. In addition, graphic communications personnel should know how to handle and prepare illustrations and photographs, prepare computer page layouts, calculate paper stock and ink needs, and cost out materials and services.

There are over 700 practical mathematical problems relating to graphic communications in this textbook. By completing these problems, students will gain a solid understanding of the fundamental principles of mathematics. In addition, the majority of the 700 plus problems have been created as story problems. Because of this, the graphic communications student must analyze and think through each problem. In other words, this textbook of practical mathematical problems is more than just performing calculations. The content of this textbook has been written to give the user the opportunity to improve upon his/her ability to determine what steps are to be taken toward solving the problems and arriving at the correct mathematical answers.

The mathematical problems in this textbook were written with the six-stage graphic communications production model in mind. This model includes the six stages of producing imaged products from design and layout through finishing and binding.

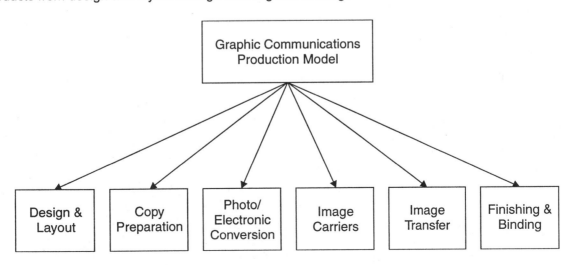

Other areas of graphic communications were also kept in mind as the mathematical problems were developed. These areas included the raw materials of ink and paper, costing procedures, sales and customer service work, and management and supervision needs. All in all, every attempt was made to include all areas of graphic communications when this textbook was being written.

ACKNOWLEDGEMENTS

This textbook revision is dedicated to my wife, B. LaVada Dennis, and to our three children, Barton Allen Dennis, Aaron Lee Dennis, and Seth Byron Dennis. Because of their wise and diligent assistance, this textbook revision became a reality. The bottom line is that this revision could not have been completed without their direct efforts and overall support of the project. Dawn Marie Dennis receives my appreciation for encouraging her husband, Barton, to devote many hours of personal time to editing and working the 700 plus practical problems in this textbook.

Several people and companies deserve recognition for their assistance in seeing this academic effort through to fruition. To Mr. Michael Stinnett, I extend special appreciation for his assistance with the introductory material in Unit 11, Combined Operations with Common Fractions. Companies and their personnel deserving appreciation include:

Hal Ehrenreich, North Central Technical College, WI
Gregory A Oxley, D. Russell Lee Career Center, OH
Walter Gunster, Tennessee Technical Center, TN
Charles J. Neal, Canadian Velley Vo-Tech, OK
Theresa R. Morlan, Wichita Area Technical College, KS
Steven E. Campeau, Milwaukee Trade-Technical High School, WI
Hamilton-Stevens Group, Inc.
Banta Corporation
Heidelberg USA, Inc.
Baumfolder Corporation
Flint Ink Corporation
Apple Computer, Inc.
Quark, Inc.
O, K & A, Inc.
Linotype-Hell Company
Hammermill Papers

Others who deserve recognition are the editors and technical personnel at ITP Delmar Publishers. There are also several people employed by companies credited in a number of figures distributed throughout the pages of this textbook who deserve special attention. Without their assistance with specific parts of this book, the publication would not have been possible.

Whole Numbers

Whole numbers are the *digits* 0 through 9 in any combination and in any number. These *numbers* include 1, 2, 3, 4, 5, 6, 7, 8, 9, and 0. To verify these numbers and their order, look at the top row of any typewriter or computer keyboard. With these ten digits, any number imaginable can be created. Most people would state that the number following 9 is ten, and that is true when counting is being done. To create two-digit numbers ten through ninety, the single numbers of 1 through 9 are combined with 0 to form 10, 20, 30, 40, 50, 60, 70, 80, and 90. To create three-digit numbers one hundred through nine hundred, the single numbers of 1 through 9 are combined with two 0s to form 100, 200, 300, and so on. The ten whole numbers are much like the twenty-six letters of the English alphabet; when numbers and letters are combined, they can become very large and powerful.

It could be interpreted that single numbers, when combined to form multiple digit numbers, are "grouped" together. One single number (1) is considered "ones"; two numbers together (10) are considered "tens"; three numbers together (100) are considered "hundreds"; four numbers together (1,000) are considered "thousands"; five numbers together (10,000) are considered "ten thousands"; six numbers together (100,000) are considered "one hundred thousands"; and seven numbers together (1,000,000) are considered "millions."

Symbols are frequently used in mathematics. They are a form of shorthand that saves space and reduces the amount of time necessary to communicate the action that is expected or required. The basic mathematical symbols include the following:

Symbol	Meaning
=	equal something as a result of previous action
+	addition of two or more number groups
-	subtraction of two number groups
×	multiplication of two or more number groups
÷	division of two number groups

 *Unit 1* ADDITION OF WHOLE NUMBERS

BASIC PRINCIPLES OF ADDING WHOLE NUMBERS

The *addition* of whole numbers one through zero in any combination and in any quantity is a straightforward task that demands attention to detail. Whole numbers should always be aligned to the right when they are being added in the traditional manner no matter how many digits they contain. For example, the following whole numbers, when aligned correctly, equal 1,735.

$$
\begin{array}{r}
15 \\
239 \\
8 \\
+\ 1,473 \\
\hline
1,735
\end{array}
$$

If these numbers are not aligned on the right side, there could be a misinterpretation and the total could be much different than intended. For example, the following whole numbers, when improperly aligned, equal 1,807.

$$
\begin{array}{r}
15 \\
239 \\
8 \\
+\ 1,473 \\
\hline
1,807
\end{array}
$$

When the single-digit number 8 is positioned one column to the left, the person doing the calculating may consider the void as a zero, thus making the number an 80 instead of an 8 as originally intended. Of course, with calculators being used to perform most mathematical calculations, this problem does not often exist. To add these four numbers with a calculator, the procedure is as follows:

$$15 + 239 + 8 + 1{,}473 = 1{,}735.$$

It is wise to use a comma to separate large numbers in groups of thousands as this helps in clarifying the number. For example, which telephone number is easier to remember? 3192732561 or (319) 273-2561? Telephone numbers are grouped to make them easier to use and remember. The same is true with numbers that are one thousand (1,000) or larger. Using commas provides separation of numbers which signifies organization, and that causes the human brain to establish a memory pattern.

PRACTICAL PROBLEMS

1. A certain number of hours are required to produce a printed job. The copy preparation artist worked 17 hours on the original copy, office corrections took 2 hours, and author's corrections took 5 hours. Photo/electronic conversion and platemaking consumed a total of 9 hours. The litho press operator used 2 hours for makeready and 7 hours for printing, while the personnel in the finishing and bindery department needed 23 hours to complete the folding, stitching, trimming, and packaging of the job. How many hours of chargeable time were required to complete this job?   _____

2. Personnel in an in-plant printing and duplicating department completed the following printed products in one day: 1,760 tickets; 860 envelopes; 2,125 business cards; 1,240 invoices; 3,600 circulars; and 1,200 labels. What was the total number of printed products completed during the single day?   _____

3. The production manager requested the following kinds of paper to be sent to the pressroom: 463 *reams* of bond, 987 reams of index, 945 reams of coated book, and 27 reams of text. (**Note:** A ream of paper contains 500 sheets.) What were the total number of reams of paper delivered to the pressroom?   _____

4. Job one took 48 hours to complete; job two took 19 hours to complete; job three required 32 hours to complete; and job four required 15 hours to complete. What were the total number of hours required to complete the four jobs?   _____

5. For a flyer advertising a musical event at a local high school, the following paper stock was used: 3,980 sheets the first day; 3,986 sheets the second day; and 4,220 sheets the third day. How many total sheets of paper were used to print the flyers?   _____

6. Employees of the Impressions Unlimited Company include 8 office workers, 13 design and layout artists, 42 copy preparation personnel, 15 photo/electronic conversion employees, 10 litho platemakers, 22 litho press operators, 34 binding and finishing personnel, 1 sales manager, 6 sales representatives, and 5 customer service representatives. What is the total number of employees employed at the Impressions Unlimited Company?   _____

7.   Employees of Central Graphics Company were very busy during the first and second weeks of a recent month. On Monday, they printed 16,150 tags; on Tuesday, they printed 17,050 tags; on Wednesday, they printed 17,500 tags; on Thursday, they printed 18,008 tags; and on Friday, they printed 17,364 tags. During the following week, they printed 4,525 more tags than the number printed the preceding week. How many tags were printed by the employees of the Central Graphics Company on the first Tuesday and first Friday and the entire following week?

_____

8.   The head accountant of the Central Graphics Company billed out four printed jobs that had been completed the previous week. The amounts were: job one, $329.50; job two, $1,645.00; job three, $893.00; and job four, $275.25. What was the total revenue from these four completed jobs?

_____

9.   Copy preparation personnel of a prepress specialty company completed a contract for typesetting four publications. Publication A contained 535 column lines of type; publication B contained 1,716 column lines of type; publication C contained 964 column lines of type; and publication D contained 2,689 column lines of type. What is the total number of column lines of type composed for publications B and C?

_____

10. Three large graphic communications companies employ the following number of people: company one, 465; company two, 238; and company three, 197. How many employees are employed at the three graphic communications companies?

_____

11. During a five-month period, the following number of copies of a popular magazine were printed, finished, and distributed: 154,598; 150,000; 162,347; 158,994; and 168,203. What was the total number of magazines printed, finished, and distributed during the five-month period?

_____

12. The number of column lines of type for publication one was 3,156 lines. For publication two, there were 1,917 column lines. For publication three, there were 5,150 column lines. How many total column lines of type were included in publications two and three?

_____

13. A litho press operator prints 13,540 letterheads on Monday; 15,794 letterheads on Tuesday; 18,947 letterheads on Wednesday; and 9,823 letterheads on Thursday. What was the total number of letterheads printed by the litho press operator during the four days? _____

14. The weekly payroll of a small screen printing company consists of the following specified amounts:

Creative artists	$ 961.55
Copy preparation computer operators	$ 1,346.16
Photo/electronic conversion operator	$ 675.45
Screen press operators	$ 2,076.92
Finishing and delivery employee	$ 524.66

What is the total weekly payroll of this screen printing company? _____

15. The manager of a county historical society was responsible for distributing several different book titles that were published by the society each year. In a recent month, the manager shipped 10,525 books during week one; 7,850 books during week two; 14,800 during week three; 16,575 during week four; and only 1,294 during the short fifth week (Figure 1–1). How many books did the county historical society manager ship during weeks one, three, and five? _____

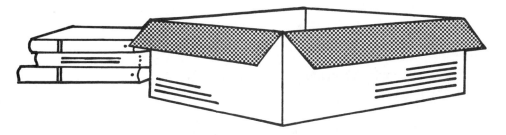

Figure 1–1 Published books must be properly packed in cartons for shipping.

Unit 2 SUBTRACTION OF WHOLE NUMBERS

BASIC PRINCIPLES OF SUBTRACTING WHOLE NUMBERS

Subtraction is the mathematical process of removing the amount of one number from another number. The result of removing a numerical amount from a given number is the *difference* or *remainder*. Only two number groups can be utilized in a subtraction procedure at one time, unlike that of addition when several numbers or number groups can be handled at one time. When subtracting numbers, it is important to keep the number groups aligned at the right side just as with addition. Also, the smaller of the two number groups being subtracted is either below or after the larger number. The following examples are used to demonstrate this statement.

Example:
$$
\begin{array}{r}
595 \\
-\,451 \\
\hline
144
\end{array}
$$
= difference or remainder

Example: $595 - 451 = 144$

It is also possible to state that a number should be subtracted from another number.

Example: Subtract 451 from 595. The result is a difference or remainder of 144.

PRACTICAL PROBLEMS

1. A paper merchant purchased a quantity of paper stock for $43,650 at a papermill and sold it to a graphic arts company for $45,000. What profit did the paper merchant gain on this sale? _____

2. There were 348 workers employed at a large graphic arts company near Chicago, Illinois, but because of a slump in company sales during the past two years, 125 workers had to be released from their jobs with the company. How many workers were left on the company payroll after the layoffs? _____

3. The owner of a "quick" print shop had 27,000 sheets (54 reams) of bond paper in stock, but the press operator used 12,500 sheets (25 reams) to print some letterhead during the morning shift. How many individual sheets of bond paper were left in the inventory to use for future jobs? _____

4. While printing two different jobs, John ran 52,598 sheets of paper stock and Mary ran 52,308 sheets of paper stock. How many more sheets of paper stock did John run than Mary?   _____

5. The first copy preparation computer operator typeset 1,002 column lines of type while the second copy preparation computer operator typeset only 376 column lines of type during the same time period. How many more column lines of type were typeset by the first computer operator than the second computer operator?   _____

6. The estimated value of a small graphic arts company building was $137,500. After the appraisal, the building was depreciated $2,550 by the appraisers. What is the present value of the building?   _____

7. A finishing person operating a paperfolder produced 46,200 *signatures* during the day shift whereas the night-shift person produced 54,030 signatures. How many more signatures were produced by the night-shift person?   _____

8. The records at inventory time showed that there were 127 reams of paper stock on hand. When the actual count was taken, there were only 119 reams available. The records were in error by how many reams?   _____

9. Estimators from two graphic communications companies submitted bids on a quantity of brochures for a well-known fishing resort in northern Wisconsin. Estimator one bid $5,239 while estimator two bid $5,457. By what amount did estimator two lose the bid?   _____

10. A copy preparation computer operator is aware that she will be required to typeset a job containing 5,227 column lines of 12-point type. During the first workday, she typesets 2,685 column lines of type (Figure 2–1). How many column lines of type will she need to typeset during the second workday to complete the job?

_____

Figure 2–1 A copy preparation computer operator must work rapidly and accurately when typesetting column lines of type. (Credit: Banta Corporation)

11. During the month of January, personnel at ABC Litho, Inc., completed 623 jobs. In January of the previous year, the same people completed 591 jobs. How many more jobs were completed this January than the previous January?

_____

12. A screen printing press operator needed to print 12,430 impressions on white, size 40 sweatshirts. Due to mechanical problems, the press operator was only able to complete 7,179 impressions on the first day, but he had two more days to complete the job. How many impressions on the sweatshirts were required in the next two days to complete the job?

_____

13. A copy preparation computer operator typeset 790 column lines of 10-point type on the first job. On a second job, the same person created an additional 955 column lines of 10 point type. How many more column lines of 10-point type were created on the second job than on the first job?   _____

14. A graphic arts company president had assets valued at $1,164,250 of which $137,500 were invested in bank stock; $127,460 were invested in mortgages; and the remaining amount was invested in the graphic arts company. How much had the company president invested in the graphic arts company?   _____

15. Two finishing and binding employees each averaged 37 hours of time during a work week. During that week, it took one employee 9 hours to complete job A, 6 hours to complete job B, and the remainder of the time to complete job C. How many hours of work time did it take to complete job C?   _____

Unit 3 MULTIPLICATION OF WHOLE NUMBERS

BASIC PRINCIPLES OF MULTIPLYING WHOLE NUMBERS

Multiplication is the procedure of increasing the size of both number groups used in the calculation. This is much the same as with addition except that only two number groups can be used at one time to perform a multiplication procedure. The following example is the traditional procedure (meaning before calculators) of performing a multiplication.

Example:

$$\begin{array}{r} 45 \\ \times\ \ 6 \\ \hline 270 \end{array}$$

The procedure involves multiplying 6 × 5 which equals 30. The zero is recorded below the line, and the 3 is carried above the 4 and will be added after the multiplication of 6 × 4 has been completed. Of course, 6 × 4 equals 24 plus 3 equals 27 which is then recorded below the line to create the number group of 270. The result of performing a multiplication of two number groups is the *product*. Most of the time, the result of the multiplication of two number groups is simply referred to as the *answer*.

When a calculator is used to perform a multiplication, the procedure involves entering either one of the number groups, pushing the × symbol key, entering the second number group, and then pushing the = key to complete the calculation. A multiplication mathematical equation is stated as follows:

$$369 \times 6 = 2,214$$

Multiplication is a simplified addition procedure in that six 369s could have been added together to achieve the same result. It should be obvious that multiplication is a much faster mathematical procedure than addition.

PRACTICAL PROBLEMS

1. The purchasing agent of a graphic arts company needed to buy six reams of 8½" x 11", 50-pound text paper. The cost of the paper was $14 per ream. What amount did the purchasing agent pay for this order? _____

2. One ream of paper contains 500 sheets, and 15 reams of paper are needed to produce a quantity of 8½" x 11" flyers. How many sheets of paper are there in the 15 reams? _____

3. An estimator determines that 346 reams of paper will be required for a printing job that is being considered by a local retail business manager. The cost of each ream of paper, according to the most recent price catalog, is $25. What will be the cost of the paper for this job?   _____

4. To produce a medium-sized garden seed catalog, 47 12" x 18" litho plates are required. The estimator determines that each plate will cost $2.35. What will be the total cost of the litho plates for producing the catalog?   _____

5. Eight letterheads can be printed at the same time on a 23" x 35" press sheet of watermarked bond paper. The litho press operator is planning to run 3,750 press sheets to complete the job. How many letterheads will be produced from this press run?   _____

6. A ream of 17" x 22" paper stock weighs 24 actual pounds, and 34 reams will be needed for producing an order of business forms. What is the total number of pounds of paper that will be used for this job?   _____

7. On average, each page of a 176-page travel guide booklet contains 216 column lines of type. How many column lines of type are in the entire booklet?   _____

8. A two-volume dictionary includes 1,980 pages of 10-point type. Each page contains 216 column lines of words and definitions. How many column lines of type are contained in the two-volume dictionary?   _____

9. It is possible to cut 110 2" x 3½" cards from one sheet of index paper stock. To complete the order for the small cards, 3,468 sheets of index paper stock will be needed. How many cards can be acquired from the 3,468 sheets of index paper?   _____

10. On average, a skilled operator can image and process a one-color litho plate containing one page in three minutes. A 64-page magazine is being produced in the Graphic Impressions Company. How many minutes will it take to produce all of the litho plates for this job?   _____

11. The labor cost of creating each full page of type, illustrations, and halftones, using standard computer software, is $46 per hour. A magazine will be produced that contains 88 pages of content. How much labor cost should be allocated for this job?

_____

12. The daily average per hour production for operators of four 19" x 25" stream-fed litho presses was 7,000 on each press. Each day consisted of eight hours. What was the average daily production of all four litho press operators?

_____

13. A skilled person operates a litho press at an average of 5,250 impressions per hour for the entire eight-hour day (Figure 3–1). How many impressions should that person be able to produce in four days?

_____

FEED SECTION

PRESS SECTION

DELIVERY SECTION

Figure 3–1   Skilled personnel are required to operate fast-running litho presses. (Credit: Hamilton-Stevens Group, Inc.)

14. At a daily newspaper, the typical copy preparation computer operator can keyboard an average of 375 column lines per hour. In five days of a 40-hour workweek, how many column lines of type can be keyboarded?

_____

15. It is estimated that 47 hours of labor will be required to complete a small, multi-page booklet. The labor rate is $11.65 per hour per person, and three people will be working on the job. How much labor cost must be charged to producing this booklet?

_____

 # Unit 4   DIVISION OF WHOLE NUMBERS

BASIC PRINCIPLES OF DIVIDING WHOLE NUMBERS

The mathematical procedure of *division* involves the dividing of one number group into two or more smaller number groups by using a smaller number group. The traditional division procedure can be stated as "divide 10 by 5." This calculation is listed by drawing the division image as follows:

$$5\overline{)10}^{\,2}$$

This procedure is often referred to as *long division* and is only accomplished when a calculator is not available. The result or *quotient* is 2 because it takes two 5s to make 10.

The mathematical method of stating the division procedure is as follows:

$$10 \div 5 = 2$$

When using a calculator, the number 10 is entered, the division symbol key is pushed, the number 5 is entered, and the = symbol key is pushed to complete the calculation and arrive at the answer or quotient.

PRACTICAL PROBLEMS

1. From one sheet of 22" x 34" paper stock, eight letterhead sheets, size 8½" x 11", can be obtained. The order from the manufacturing company calls for 15,000 letterheads. How many 22" x 34" stock sheets will be needed for the job? _____

2. A flyer has been printed and now it is time to fold the printed sheets with two parallel folds to create the six pages contained in standard flyers. The total job calls for 42,250 flyers, and five skilled paperfolder operators will be responsible for the folding. Assuming that all workers do the same amount, what will be the average number of copies folded by each person? _____

3. There are 458,500 sheets of paper stock on the shelves in the stockroom of the graphic arts company. With the understanding that a ream constitutes 500 sheets, how many reams are there in the 458,500 sheets of stock paper? _____

4. It is possible to purchase 25 reams of 25" x 38" coated book paper for $975. What is the cost of the paper per ream?   _____

5. There are 23 people employed at the Jones Embroidery and Sign Company located in a large southern city. Their benefits administrator negotiates a health care package that amounts to $13,754 for the coming year. To what extent does each employee benefit?   _____

6. A small litho press was purchased by a job-shop proprietor for $26,470, including interest (Figure 4–1). Equal monthly payments are to be made over a 27-month period. What is the amount payable each month?   _____

Figure 4–1   Small litho presses are used to produce a large percentage of the printed products that are used in business and industry. (Credit: Heidelberg USA, Inc.)

7. In the Graphics Unlimited Company, 29,925 advertising flyers were printed in a seven-hour time period. What was the average number of flyers printed per hour?   _____

8. A booklet contains 2,360 lines of type and there are 40 lines on each page. How many printed pages are there in the booklet?   _____

9. In one month, a commercial printer specializing in producing business cards sold 369 orders for a total of $6,059. What is the average selling price for each order?

_____

10. An estimate has been prepared for a customer's catalog. It was determined that 42 large metal litho plates will be needed to complete the job. The total cost for the large metal litho plates will be $504. What is the cost for each litho plate?

_____

11. A paper cutter operator cuts four equal lengths from a sheet that measures 36" long. What is the length of each piece?

_____

12. The payroll of a graphic arts company includes the following information for a 5-day, 40-hour week:

3	Copy preparation computer operators	$1,620
1	Litho platemaker	$ 550
1	Litho press operator	$ 635
1	Binding and finishing person	$ 495
1	Shipping department clerk	$ 400

What is the hourly wage for each of the seven company employees?

_____

13. A batch of 10,000 printed and folded pamphlets will be mailed. The total weight, before wrapping them into bundles, is 2,500 pounds or 40,000 ounces. What is the weight in ounces of each pamphlet?

_____

14. A copy preparation computer operator is capable of typesetting an average of 150 column lines per hour. How much time will it take for the computer operator to typeset a job, assuming the average production rate is met, that contains 5,100 column lines of type?

_____

15. For some special watermarked, rag content bond paper, the cost is $8.72 per ream. The customer has $375 budgeted for paper for a printed job that is being planned. How many reams of paper will the customer be able to purchase?

_____

Unit 5   COMBINED OPERATIONS WITH WHOLE NUMBERS

BASIC PRINCIPLES OF COMBINING OPERATIONS WITH WHOLE NUMBERS

Whole numbers are, without doubt, the most important of all the numbers that can be assembled. As indicated in the opening paragraph of this section, the numbers 0 through 9 are used in any combination to form the smallest to the largest numbers ever conceived. Whole numbers can be and are used as singles or "ones," doubles or "tens," triples or "hundreds," quadruples or "thousands," and beyond.

Knowing when to add, subtract, multiply, and/or divide whole numbers is very important in our daily lives. For example, it is very valuable to know when these mathematical functions should be completed in balancing your checkbook or in calculating your state and federal income taxes. It is true that devices ranging from basic calculators to high-powered computers are being widely used for these functions, but knowing how to verify the "electronic" transactions is reason enough to know when certain basic whole number calculations should or should not be made. The problems in this unit have been designed to challenge your whole number calculation skills and your ability to determine when to add, subtract, multiply, and/or divide.

PRACTICAL PROBLEMS

1. John, a graphic arts customer service representative, made it a point to balance his personal checkbook each Saturday morning. Last Saturday morning, his checkbook balance was $543. This past week he wrote five checks for the amounts of $26, $113, $4, $76, and $145. What was the balance of his checkbook account after writing these checks?   _____

2. The owner of a binding and finishing company has been renting the building in which the company is located. She has paid an average of $4,350 per month for the rent, and the rent has been consistent for the past two years. How much money has been spent for building rent for the two-year period?   _____

3. The manager of the shipping and delivery department of a large publication printer was required to complete a report each month. There were five delivery vans and two 18-wheel semi-trailer trucks in the fleet of vehicles. Each of the seven vehicles was always filled with fuel at the end of the last working day of each month so accurate records could be kept. Each van was filled with gas five times each month plus the filling on the last working day. Also, the 18-wheel semi-trailer trucks were filled with diesel seven times each month plus the filling on the last working day. How many total fuel (gas and diesel) fillings were there among the fleet of vans and trucks each month?

    _____

4. Many gallons of fuel were burned in the seven vehicles presented in problem 3. In fact, the following number of gallons of gas were used in the five delivery vans for a recent month: 105, 97, 111, 137, and 80. What was the average number of gallons of gas burned in each of the five delivery vans for the month?

    _____

5. Gas is expensive, but it must be purchased for use in the five delivery vans as presented in problem 3. The average price of the gas for this past month was $2 per gallon. What was the total cost of the gas for the five delivery vans for the past month?

    _____

6. As everyone knows, there are 366 days in a leap year. There are 52 Saturdays and 52 Sundays on which many graphic arts companies are not open for business. In addition, there are a minimum of seven holidays that are observed each year on which most graphic arts companies are closed. How many work days are there in a typical leap year after deleting the Saturdays, Sundays, and holidays?

    _____

7. There are 52 computers distributed throughout a medium-sized, daily newspaper, but there are only 29 printers for the computers. A maximum of two printers are shared per computer, but each of the computers is attached directly to a printer. How many of the computers are attached to only one printer?

    _____

8. The publishers of a large textbook company have 246 active book titles listed in their annual sales catalog. The goal of the sales department personnel is to sell an average of 5,000 books per title per year. How many total copies of books must be sold each year to meet this sales goal?

    _____

9. A digital press operator imaged 6,784 sheets of paper for eight customers. What was the average length of run for each customer? _____

10. Eighteen graphic communications executives and educators are members of an active graphic communications advisory committee. Some of the members live and work in the local area, but most of them live and work out of the city or even out of the state (Figure 5–1). The total round-trip miles that the eighteen committee members travel for an advisory meeting equals 9,090. What is the average number of miles, round-trip, that each committee member travels for a meeting? _____

Figure 5–1 Members of an active graphic communications advisory committee took time from their semi-annual meeting to pose for this photograph.

11. Lunches and refreshments for a meeting of the graphic communications advisory committee average seven dollars each. For a regularly scheduled meeting, 12 committee members and 10 guests are generally present. How much does it cost to provide the lunches and refreshments for the committee members and guests for a meeting?   _____

12. There are 75 declared majors and 18 declared minors in the graphic communications curriculum at a regional university. Of the total number of majors and minors, there are 24 seniors and 29 juniors. How many students of this total group are freshmen and sophomores?   _____

13. The employees of a 24-hour copy service business made 5,280 photocopies on a particular Saturday. Determine the average hourly output of photocopies for the 24-hour period.   _____

14. A photographer, working for a catalog publisher, used 15 rolls of film with 36 exposures each to photograph products that will be displayed in the Christmas edition of a department store catalog. The photographer went on to use 6 rolls of 36-exposure film for a toy store catalog. How many more exposures were taken for the department store catalog than the toy store catalog?   _____

15. There are 99 counties in an average-sized state in the central United States. Within this state, there are 1,287 graphic communications-related companies. What is the average number of graphic communications companies per county?   _____

 # Unit 6   *ROMAN NUMERALS*

BASIC PRINCIPLES OF ROMAN NUMERALS

People working in the graphic communications industry should be familiar with *Roman numerals*. They are sometimes used on title pages, for chapter headings, and to specify volume numbers. Paper manufacturing company personnel and paper supply company personnel often use the Roman numerals C, D, and M after the weight or the unit cost of paper stock to indicate the weight or cost per 100, 500, or 1,000 sheets of paper (Figure 6–1). `

TEXT and TEXT COVER							
HOWARD LINEN COVER							
				PRICE PER CWT.			
				BKN. CTN.	1 CTN.	4 CTN.	16 CTN.
			White	187.70	107.25	92.95	85.80
			Light Colors	196.45	112.25	97.30	89.80
			Dark Colors	202.05	115.45	100.05	92.35
Basis 20 × 26	Size	M Wt.	Ctn. Pack	PRICE PER 1000 SHEETS			
				BKN. CTN.	1 CTN.	4 CTN.	16 CTN.
WHITE, WARM WHITE							
65	23 × 35	201	750	377.28	215.57	186.83	172.46
80	23 × 35	248	500	465.50	265.98	230.52	212.78

Figure 6–1   Roman numerals, although not frequently used, are found on some paper packaging labels and in paper catalog price listings.

With this in mind, it is important to know the numerical values established for seven capital letters of the alphabet. They are listed as follows:

letter I = 1   letter C = 100

letter V = 5   letter D = 500

letter X = 10   letter M = 1,000

letter L = 50

Expressing Arabic Numerals as Roman Numerals and Vice Versa

The basic principles of addition are used in the Roman system of *enumeration*. When addition is used, this means that separate letters are added (starting with the largest number first) to determine the *Arabic number*. The following examples help to demonstrate the additive principle.

Examples: VI   = 6   XI   = 11
XV   = 15   LX   = 60
LXX = 70   LXXX = 80

The subtractive principle is also used in six specific instances. When subtraction is used, a lesser value symbol is placed before a larger value symbol. This indicates that the smaller number is subtracted from the larger number. The following examples are the six instances in which the subtractive principle is used.

Examples: IV   5 – 1   = 4   IX   10 – 1   = 9
XL   50 – 10   = 40   XC   100 – 10   = 90
CD   500 – 100 = 400   CM   1,000 – 100 = 900

Both additive and subtractive principles are used together to form Roman numerals that translate to standard Arabic numerals. Several examples are listed below.

Examples: XIV   = 14   XIX   = 19
XXIV   = 24   XXIX   = 29
XXXIV = 34   XXXIX = 39

The following examples illustrate how the additive and subtractive principles are used to form Roman numerals.

Examples: 1. 55   = 50 + 5 or L + V   = LV
2. 1944   = 1,000 + 900 + 40 + 4 or M + CM + XL + IV   = MCMXLIV
3. CLV   = C + L + V or 100 + 50 + 5   = 155
4. CDXXIV = CD +X + X + IV or 500 – 100 + 10 + 10 + 5 – 1 = 424

Special Marks and Uses

When a bar is placed over a letter or letters, the amount must be multiplied by 1,000. For example, $\overline{XX}$ = 20,000 instead of only 20; and $\overline{L}$ = 50,000 instead of only 50.

Lowercase Roman numerals are sometimes used to number introductory pages of a publication such as this and other books. For example, page three of the introductory pages of a book may be written as iii, page four as iv, page five as v, page eight as viii, and so on.

PRACTICAL PROBLEMS

A.   Write the following Arabic numerals as Roman numerals.

1.	19	_____	6.	112	_____
2.	14	_____	7.	549	_____
3.	25	_____	8.	494	_____
4.	69	_____	9.	2,494	_____
5.	96	_____	10.	1,979	_____

B.   Write the following Roman numerals as Arabic numerals.

11.	XX	_____	16.	DCXLVII	_____
12.	XII	_____	17.	DIV	_____
13.	CCCV	_____	18.	MMCLIX	_____
14.	CVIII	_____	19.	CMXCIX	_____
15.	MCCCL	_____	20.	DCCXCVII	_____

C.   Write the following Arabic numerals as Roman numerals.

21.	6,000	_____	23.	75,542	_____
22.	53,647	_____	24.	653,947	_____

D.   Write the following Roman numerals as Arabic numerals.

25.	vii	_____	28.	$\overline{\text{LX}}$	_____
26.	iii	_____	29.	$\overline{\text{CXXX}}$	_____
27.	xvi	_____	30.	$\overline{\text{CXLVII}}$CXXII	_____

Common Fractions

Common fractions, such as ½, ¼, ⅛, and ¹⁄₁₆, represent partial whole numbers. There are many *fraction* numbers; in fact, there are about as many fraction numbers as there are people willing and needing to work with fractions. Fractions are typically written as they are in the first sentence ½, 1/2 or as $\frac{1}{2}$. With computer software programs, the first two formats are the most common.

There are three parts to a fraction. The first or top number is called the *numerator*, the line, either diagonal or horizontal, separating the two numbers is called the *fraction bar*, and the second or bottom number is called the *denominator*. The denominator of a fraction denotes the number of equal pieces or parts into which a whole number or object has been divided. In turn, the numerator denotes the number of pieces or parts that have been consumed. For example, with the fraction ¾, the measurement or the object itself has been divided into four parts and three of the parts have been used to suggest a specific measurement or size of object.

Adding, subtracting, multiplying, and dividing fractions that do not contain common (or the same) denominators is a difficult to impossible task. It is necessary to first make both denominators the same. This is accomplished by multiplying the numerator and denominator of each fraction by whole numbers that will result in the same denominator for both fractions.

Example: Add ½ and ¼.
    ½ × 2 = ²⁄₄
    ²⁄₄ + ¼ = ¾

Note that the numerators are added, but the denominator of each fraction is kept in its single form.

There is no absolute rule as to which whole number should be used to create common denominator fractions. This sometimes takes some experimentation while working with fractions with uncommon denominators.

Unit 7   ADDITION OF COMMON FRACTIONS

BASIC PRINCIPLES OF ADDING COMMON FRACTIONS

Common fractions, those with the same denominator, are easy to add. The numerators of the two or more fractions are added together, and the common denominator is retained as is.

Example: $\frac{1}{16} + \frac{5}{16} + \frac{4}{16} = \frac{10}{16}$

This new fraction can now be reduced in number size of both the numerator and denominator by dividing both fraction numbers by the same whole number. In this case, 10 can be divided by 2 to equal 5, and 16 can be divided by 2 to equal 8; thus, the reduced fraction is $\frac{5}{8}$. This is the smallest numerical amount for both the numerator and denominator because no other common whole number can be divided into the 5 and 8.

Uncommon fractions that are to be added together (for example, $\frac{1}{8} + \frac{1}{4} + \frac{1}{3} = ?$) must receive a conversion of the denominators so all of them are the same. One method is to multiply the three denominators by each other:

$$8 \times 4 \times 3 = 96$$

thus, $\frac{1}{8}$ would equal $\frac{12}{96}$, $\frac{1}{4}$ would equal $\frac{24}{96}$, and $\frac{1}{3}$ would equal $\frac{32}{96}$ for a total of $\frac{68}{96}$. The numerator is acquired by dividing the common denominator of all three fractions by the numerator of each fraction. Using the example of $\frac{1}{8}$:

$$96 \div 8 = 12, \text{ thus } \frac{1}{8} = \frac{12}{96}$$

These numbers work, but they are rather large and cumbersome. It is wise to determine whether a smaller denominator could be used to achieve the same results. In this example, the number 24 is smaller and is divisible by all three denominators. The fractions would then be added as follows after both the numerator and denominator are multiplied by the whole number of 24:

$$\frac{1}{8} = \frac{3}{24}, \frac{1}{4} = \frac{6}{24}, \text{ and } \frac{1}{3} = \frac{8}{24}$$

$$\frac{3}{24} + \frac{6}{24} + \frac{8}{24} = \frac{17}{24}$$

Sometimes, the numerator becomes larger than the denominator when a series of fractions are added together.

Example: $\frac{5}{8} + \frac{3}{8} + \frac{7}{8} = \frac{15}{8}$

24

This situation or condition is called an *improper fraction*. When the numerator is larger than the denominator, this indicates that the fraction is larger than at least one whole number, and in most situations, the improper fraction should be converted to a *mixed number*. An improper fraction can easily be converted to a mixed number by simple division:

$$15 \div 8 = 1\tfrac{7}{8}$$

A mixed number is one that contains a whole number and a fraction for a given measurement.

PRACTICAL PROBLEMS

1. A job ticket for a sales leaflet contained the following prepress work information: artwork, 2½ hours; copy preparation, 3½ hours; photo/electronic conversion, ½ hour; and color proofing, 1½ hours. What was the total work time recorded on this job ticket? _____

2. The litho press department supervisor ordered the following quantities of ink: warm red, 5¼ pounds; fast-dry black, 10¾ pounds; *cyan*, 7⅝ pounds; and *magenta*, 2½ pounds. How many pounds of ink were ordered by the department supervisor? _____

3. An estimator calculates that the following amounts of paper will be needed for these jobs: job A, 21½ reams of book paper; job B, 12⅜ reams of #4 sulfite bond paper; and job C, 7³⁄₁₆ reams of 25% rag bond paper. What is the total number of reams of paper required for the three jobs? _____

4. A section of a parts catalog included some tabular matter with the following column widths: column one, 5¼ *picas*; column two, 3½ picas; column three, 4¾ picas; column four, 8 picas; and column five, 5½ picas (Figure 7–1). What was the total number of picas consumed in the five column widths? _____

CATALOG NUMBER	COLOR	ITEM	DESCRIPTION	PRICE
AN 321	RED	201	SOLID STATE	$29.95 EA
AN 322	WHITE	202	SOLID STATE	$29.95 EA
AN 323	BL	203	STATE	$29
AN 32				

Figure 7–1   An example of a five-column parts catalog.

5. Prior to sitting down at the computer keyboard, a copy preparation person prepared a layout for a sixteen-page booklet that will contain one horizontal fold in the middle and two parallel folds to create the sixteen pages. The folder will be printed on both sides; thus, each side of the sheet will contain eight pages of exacting dimensions (Figure 7–2). The horizontal dimensions of the eight pages and margins beginning from the left are listed as follows: ⅛", 5½", ⅛", ⅛", 5½", ⅛", ⅛", 5½", ⅛", ⅛", 5½", and ⅛". What is the total width of the sheet that will be needed for this small booklet? _____

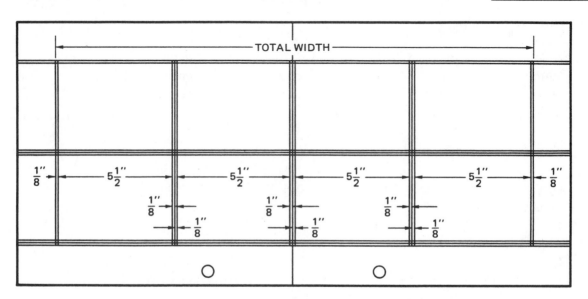

Figure 7–2   The eight-page layout of a sixteen-page booklet.

6. For the booklet job referred to in problem 5, the layout artist has planned a *bleed* to be printed on one of the pages next to the *gripper margin*. The copy preparation person rules out the following vertical dimensions: ⅜", ⅛", 4¼", ⅛", ⅛", 4¼", and ⅛". What is the total depth or vertical height of the sheet needed for this job? _____

7. For a more economical press run, saving both labor and material, several printing jobs will be *ganged* together. The widths of the four jobs are listed as follows: job one, 3⅛"; job two, 4³⁄₁₆"; job three, 7¹³⁄₁₆"; and job four, 10⅜". What will be the total amount of paper consumed, width wise, for these four jobs? _____

8. To receive freight prepaid on a quantity of paper from the paper supply company, a graphic arts company buyer was required to order 500 pounds of paper or more. The buyer placed an order for 127½ pounds of ledger stock, 60¼ pounds of #4 sulfite bond stock, 75¾ pounds of cover stock, 220⅜ pounds of index stock, and 123⅞ pounds of 10-point board stock. The buyer hoped these amounts of paper would equal the total needed for the prepaid freight qualification. What was the total number of pounds of paper stock ordered?

_____

9. In the embroidery department of the Creative Textile Company, three different shifts of employees worked on the same job. The first-shift employee worked 6¼ hours, the second-shift employee worked 7⅛ hours, and the third-shift employee worked 5⅜ hours. What was the total number of hours spent on the job by employees of the three shifts?

_____

10. During a typical workweek, the employees in the flexography plate department worked on several jobs including labels for glass bottles, wrapping paper, plastic bags, and some paper grocery sacks. Their productive time was recorded as follows: Monday, 6³⁄₁₀ hours; Tuesday, 7⁵⁄₁₀ hours; Wednesday, 6⁴⁄₁₀ hours; Thursday, 5⁹⁄₁₀ hours; and Friday, 7¹⁄₁₀ hours. What was their total productive time for the entire week?

_____

11. In preparing a comprehensive layout for a two-page layout, the graphic artist measured the following dimensions for the lines that had to be drawn (Figure 7–3): left margin, ⅝ inch; type area, 4¾ inches; gutter, 1¼ inches; photo, 2½ inches; white space, ⅛ inch; type area, 2⅛ inches; and right margin, ⅝ inch. What is the total width of the two-page layout?   _____

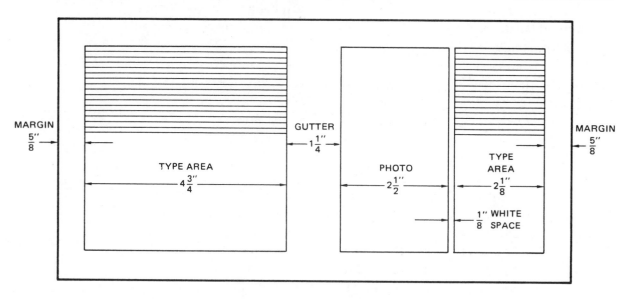

Figure 7–3 A two-page comprehensive layout containing critical positioning measurements of the planned content.

12. During the periodic warehouse inventory of the All-Business Envelope Company, the employees recorded the following information: 4½ cases of #6¾ white envelopes; 7¾ cases of #10 white envelopes; 1¾ cases of #9 return blue envelopes; 2¼ cases of #10 white window envelopes; and 3¼ cases of ivory colored A-2 envelopes. What is the total number of white envelopes in inventory?   _____

13. A sales representative recorded the following times on a call record sheet: ½ hour travel and 1¼ hours with customer A; ¼ hour travel and 1⅜ hours for lunch with customer B; ¾ hour travel and ¼ hour with customer C; and ¾ hour travel to home. What was the total amount of time directly spent with the three customers?   _____

14. The layout of a newspaper editorial page was being planned for the Friday edition. The editor decided on a five-column format for the page. Each column will be 2⅛ inches wide; the distance between each column will be ⅛ inch; and each margin will be ¾ inch. What will be the total width of the newspaper page? _____

15. A three-column advertisement has been sold that will appear in the newspaper referred to in problem 14. The editor uses the same column measurements as those specified for the newspaper layout (Figure 7–4). What is the total width of this advertisement? _____

Figure 7–4 The rough layout of a five-column newspaper page showing the position of the planned three-column advertisement.

Unit 8   SUBTRACTION OF COMMON FRACTIONS

BASIC PRINCIPLES OF SUBTRACTING COMMON FRACTIONS

The same basic principles of handling fractions when they are being added should be followed when there is need to subtract fractions. When fractions contain the same denominator, the numerators are subtracted to obtain the difference between the two fractions.

Example: $12/16 - 7/16 = 5/16$

When fractions must be subtracted that have different denominators, it is necessary to find common denominators for the two fractions, and then the difference between the fractions can be calculated.

When one or more mixed numbers are involved, it normally works best to convert each mixed number to an improper fraction and then the subtraction calculation can be made.

Example: Subtract $7/8$ from $1 5/8$.
First, convert $1 5/8$ to $13/8$
$13/8 - 7/8 = 6/8$, which is reduced to $3/4$

Note: Fractions should generally be reduced to the lowest terms possible so they can be remembered and understood better.

The procedure used to reduce $6/8$ to $3/4$ in the example was to select a whole number that could be divided into both the numerator and denominator and achieve numbers that were even. The common number was 2, thus $6 \div 2 = 3$ and $8 \div 2 = 4$. Again, there is no hard-and-fast rule or procedure for arriving at the lowest terms for a fraction except to look closely and see if it is possible to identify a common number that can be divided evenly into both the numerator and denominator.

PRACTICAL PROBLEMS

1. A standard typesetting job was keyboarded by James on the company computer system, and it took him $3 3/4$ fewer hours than it took his coworker Linda. She, being new to this type of composition, took $10 1/4$ hours to keyboard the job. How many hours did it take James to typeset the job?   _____

2. A typesetting job was set 7½ inches (approximately 45 picas) wide and printed on a sheet 9¼ inches wide with a ¾-inch margin on the left side of the page (Figure 8–1). How much space remained for the right margin of the page? _____

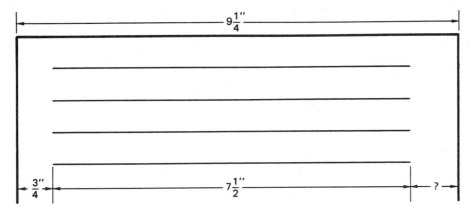

Figure 8–1 A rough layout of a page containing a single column and the standard side margins.

3. A high-speed digital printer operator noted that there were 8¾ reams of light blue book paper left in the stockroom. She had a job to duplicate that would take 7⅓ reams of the book paper. How many reams of light blue book paper were left after completing this job? _____

4. A flexographic press operator obtained 7⅝ pounds of emerald blue ink from the supervisor in the ink supply room. He used 2¾ pounds of the ink for one job and planned to use the remainder of the ink for the second job that he had to print. How much emerald blue ink was remaining after the flexographic press operator completed the first job? _____

5. When the regular monthly inventory was conducted, the inventory supervisor counted 27¾ reams of canary, 65-pound cover paper stock. According to the computer printout, there should have been 28½ reams of this paper remaining. What was the difference between the inventory taken by the supervisor and that reported on the computer printout? _____

6. The estimator of the Full-Color House prepress company calculated that the museum two-color brochure job would take 17³⁄₁₀ hours to complete. When the time sheet arrived back in the estimator's office, the actual time spent on the job was 15⁶⁄₁₀ hours. How many fewer hours did it take the designers and copy preparation personnel to complete the job than was estimated?

7. A one-page flyer is being prepared by two members of the company art department. The customer wants the flyer to fit into a #10 envelope that measures 4⅛ inches by 9⁷⁄₁₆ inches (Figure 8–2). It was quickly decided that the flyer should be 9⅛ inches long, but there was some indecision about how wide it should be. Because of the thickness of the paper, it was decided that it must be ³⁄₁₆ inch less than the height of the #10 envelope. The flyer will be sized 9⅛ inches by what measurement?

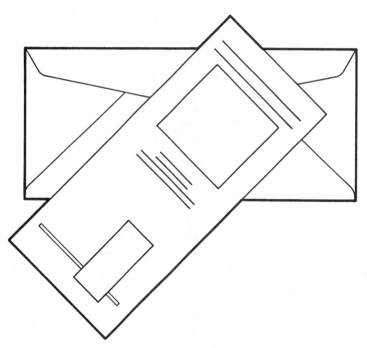

Figure 8–2   A brochure should be sized so it can be easily placed in a #10 business envelope.

8. A graphic communications company president sent three of her plant personnel to a seminar in a distant city. It was estimated that under normal driving conditions, the trip would take 3¼ hours. The traffic was much lighter than expected, and the trip only took 2¾ hours. How much time was saved on their drive to the distant city?

    _____

9. A paper cutter operator cut a 23" (width) x 35" (length) sheet of text paper stock into the following lengths: 8½ inches, 8¼ inches, 8⅝ inches. What is the length of paper that remains?

    _____

10. A total of 14⁹⁄₁₀ hours are needed to complete a glass-etching run. Members of the day shift start the job and record 7½ hours on the job ticket time sheet. How much time remains for members of each of the following two shifts to equally complete the job?

    _____

11. The prepress proofs must be mailed within 6³⁄₁₀ hours. The copy preparation personnel need 3⁶⁄₁₀ hours to finalize the copy and prepare the film negatives via the on-line imagesetter. How much time will remain for the color proofing department personnel to prepare a set of full-color transfer proofs?

    _____

12. Because of some machinery breakdowns, a job that normally takes 36¾ hours to complete actually took 42⁷⁄₁₀ hours to complete even though all employees were working as rapidly as possible. By how much time did the employees miss the usual number of hours required for this job?

    _____

13. Inventory control is an important factor in any business, especially in a close profit margin graphic communications business. The computer inventory records indicate that there are 27¾ pounds of black toner on hand in the storeroom. A new, but very reliable, employee can locate only 24⁷⁄₁₆ pounds of the toner. How many pounds short is the actual supply of black toner?

    _____

14.   A customer arrived at the Fifth Avenue Quick Print Company promptly at 8:00 A.M. on Wednesday of a busy week. The customer needed three reports printed from prepared plate-ready copy for a 2:30 P.M. meeting that same day. The press operator stated that it would take ¾ hour for the first report, 1¼ hours for the second report, and 1½ hours for the third report. In addition, the binding and finishing personnel stated that it would take 2 hours to complete stitching and drilling holes in the three reports. By what margin of time did the personnel of the Fifth Avenue Quick Print Company believe that they could complete the job?

                                              _____

15.   A copy preparation computer operator was required to gang up several jobs for a long litho press run. The widths of three jobs were listed as follows: job A, 6¼ inches; job B, 8¾ inches; and job C, 6¹⁵⁄₁₆ inches. The press sheets cut from the stock sheets were 23 inches wide (Figure 8–3). How much trim was left?

                                              _____

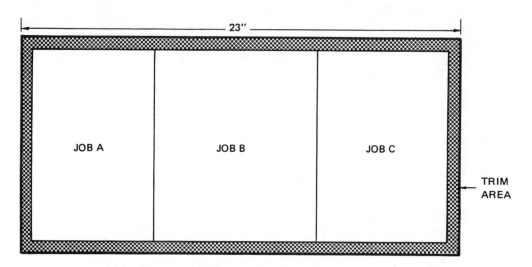

Figure 8–3   Ganging jobs on one sheet saves both labor and paper stock.

6. At the Nelson Publication Gravure Company, there are four skilled press operators who average 7½ hours of work time per day. The workweek generally consists of the standard five week days, plus one-half day on Saturday, for a total of 5½ days per week. What is the total number of hours worked by the four gravure press operators during an average workweek?   _____

7. Type, photographs, and illustrations constitute a standard newspaper page of five columns with each column 2³⁄₁₆ inches wide. Each of the five columns is separated by ⅛ inch, and both the right and left margins are ¾ inch wide. How many inches wide is the standard newspaper page?   _____

8. Five tickets will be prepared on one sheet by the copy preparation computer operator (Figure 9–2). Each ticket measures 2⅛ inches deep, and ⅜ inch must be planned for the litho press gripper margin. How long must the index paper stock be to accommodate the five tickets and gripper margin?   _____

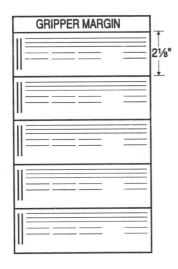

Figure 9–2 Small tickets can be handled much easier when they are grouped on one sheet by the copy preparation computer operator.

9. A case-bound technology-based book weighs 1¼ pounds. There are 1,570 of these books in inventory at the Best Binding and Finishing Company. What is the total weight of these technology books?   _____

10. For each batch of magazines, 2⅜ rolls of 50-pound book paper were used. To complete the contract, 61½ batches of magazines had to be produced. How many rolls of 50-pound book paper were used for this job?   _____

11. At the Label Flexography Company, black flexo ink was being used at the rate of 8⁷⁄₁₆ pounds per hour on one press and 7¼ pounds per hour on a second press. On Tuesday of the one week, each press was operated for 6⁷⁄₁₀ hours. How many pounds of ink were used on that Tuesday?   _____

12. For a specific kind of type, a line of 10-point type contains an average of 11 words. The leaflet being designed by the layout artist needs 175 lines to cover all of the planned content. How many words are in the leaflet?   _____

13. The completion of the binding and finishing of a large wholesale catalog required three shifts of employees working 7½ hours per shift for four days. What is the total number of hours charged to this job?   _____

14. A computer operator in the Stils Design Company worked 8³⁄₁₀ hours a day for five days last week. The same computer operator then came into work over the weekend and recorded 5⁹⁄₁₀ hours of additional time. How many hours were recorded on the computer operator's time card for the entire week?   _____

15. A ream of special paper weighs 22½ pounds, and there is one available skid of paper containing 40½ reams. How many pounds of this special paper are available?   _____

Unit 10   DIVISION OF COMMON FRACTIONS

BASIC PRINCIPLES OF DIVIDING COMMON FRACTIONS

The division of fractions involves a procedural step that is not common to adding, subtracting, and multiplying fractions. This step involves changing the positions of the numerator and denominator on the *divisor* fraction. This procedural step is referred to as the *reciprocal* of a fraction.

Example: The reciprocal of $7/8$ is $8/7$

The number 8 becomes the numerator and the number 7 becomes the denominator. This is a simple procedure, but it is critical when dividing fractions.

Once the reciprocal of the divisor fraction has been determined, the next step is to multiply the fractions using the procedure described in Unit 9, Multiplication of Common Fractions. Following this multiplication step, the fraction should be reduced and then, if necessary, changed into a mixed number.

Example 1: $3/4 \div 2/3 = ?$
$3/4 \div 3/2 = ?$ The divisor fraction is written in reciprocal form.
$3/4 \times 3/2 = 9/8$ The numerators are multiplied and the denominators are multiplied.

The $9/8$ answer cannot be reduced, but it can be changed into a mixed number of $1\frac{1}{8}$.

Example 2: $13/16 \div 3/8 = ?$
$13/16 \div 8/3 = ?$ Write the divisor fraction in reciprocal form.
$13/16 \times 8/3 = 104/48$ Multiply the numerators and denominators.
$104/48 \div 8 = 13/6$ Reduce the fraction by dividing both the numerator and denominator.
$13/6 = 2.1667 = 2\frac{17}{100}$ Convert to a mixed number.

Note: The $17/100$ fraction cannot be reduced, but for general understanding, it is between the fractions $1/6$ and $1/5$.

Two other basic rules to remember when dividing with fractions include the following: (1) when mixed numbers are involved, change them to uncommon fractions, and (2) when dividing a whole number by a fraction, make the whole number a fraction as described in Unit 9, Multiplication of Common Fractions. Once these two steps are completed, the division procedure can be completed in a straightforward manner.

PRACTICAL PROBLEMS

1. A computer word processing specialist completed the keyboarding of a manuscript containing 7,650 individual type characters in 4¼ hours. How many type characters were typeset per hour?

    _____

2. A line of type measures 25 picas long and contains 62½ type characters. How many type characters are there in each pica?

    _____

3. A litho press operator recently produced 69,375 impressions during a 7½-hour workday. On the average, how many impressions were produced per hour by the press operator?

    _____

4. The employees of a wholesale paper company cleared out some of their odd-lot paper stock. They decided to dispose of one of the skids that contained 750¾ pounds of odd-lot paper. After checking, they found that each ream of the paper weighed 20 pounds. How many reams of paper stock were contained on the skid of paper?

    _____

5. Paper-folding machines are sometimes rated according to the number of inches of paper that can be run through per hour rather than the usual sheets-per-hour rating (Figure 10–1). While operating one of the several graphic arts company's paperfolders, a skilled operator logged 106,562½ inches of paper in 3⅞ hours. How many inches per hour was the operator able to run through the paperfolder?

    _____

Figure 10–1 Paper folding machines are available in many sizes and designs, but most can be operated at high rates of speeds. (Credit: Baumfolder Corporation)

6. The supervisor of the estimating department of the Highland Graphic Technology Company prepared an estimate for 85,000 booklets. She calculated that it would take 12²/₁₀ hours of litho press time to complete the job. How many impressions per hour must the press operator average to complete the job according to the time specified by the estimator? _____

7. A technician in the photo/electronic conversion department was responsible for running the imagesetting machine. On average, it was possible for him to produce 152 line negatives during each 7⁶/₁₀-hour shift. How many line negatives could he produce per hour? _____

8. When costing out a pressure-sensitive label job, it was found that 22¾ pounds of ink were used by the flexography label press. The total cost of the ink was $182. What was the cost of the ink per pound? _____

9. Book paper is generally available in sheet sizes that measure 25" × 38". A job to be cut from this stock requires a sheet that measures 9⁷/₁₆" × 12⅜". How many 9⁷/₁₆-inch sheets can be cut from the 38-inch length? _____

10.   Review the information that was given in problem 9. How many 12⅜-inch sheets can be cut from the 25-inch measurement?

_____

11.   Review the information given and requested in problems 9 and 10. How much paper was remaining on the 25-inch measurement after the 12⅜-inch sheets were removed?

_____

12.   A manuscript containing 10,576 words was submitted to the publisher by a well-known author. The typographer, designing the body of the book where this manuscript will be used, determined that each page of type should average 210½ words. With that information in mind, how many pages of type will be contained in the manuscript?

_____

13.   A paper cutter operator was required to cut 5¼" × 7½" cards from index paper stock size 25½" × 30½". How many 5¼-inch strips for the cards could be cut from the 25½-inch length of paper stock?

_____

14.   Review the information found in problem 13. How much paper would remain of the 30½-inch length if the maximum number of 5¼-inch cards were cut from this measurement?

_____

15.   The typesetting specialist was asked to typeset tabular matter on a standard computer with a special text software program. The job called for 10 columns of type to fit equally into 37½ picas (Figure 10–2). How wide did each column need to be?

_____

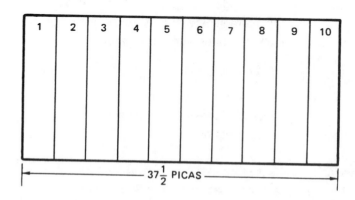

Figure 10–2   For this job, each of the ten columns of type had to be of equal size.

Unit 11 COMBINED OPERATIONS WITH COMMON FRACTIONS

BASIC PRINCIPLES OF COMBINING OPERATIONS WITH COMMON FRACTIONS

As stated in the Section 2 introduction, fractions represent a portion or part of a given amount. For example, ½ of an amount, such as ½ of 16 houses or ½ of a gallon container, immediately suggests that there are two parts to equal the whole. The denominator always represents the total number of parts, and the numerator indicates the number of parts that are being considered. With ¼, for example, one of the four parts is represented, whereas with ¾, three of the four parts are being considered.

Halving Common Fractions

A quick, easy, and accurate method of halving common fractions is presented as follows. This procedure can save considerable time when working with common fractions.

Example 1: Finding half of the common fraction ⅞.
Step 1. Leave the numerator the same: 7.
Step 2. Double the denominator: 8 × 2 = 16.
The answer is ⁷⁄₁₆, which is half of ⅞.

Example 2: Finding half of a mixed number fraction with an even whole number: 8⅜.
Step 1. Divide the whole number by 2: 8 ÷ 2 = 4
Step 2. Leave the numerator of the fraction the same: 3.
Step 3. Double the denominator: 8 × 2 = 16.
The answer is 4³⁄₁₆.

Example 3: Finding half of a mixed number fraction with an uneven whole number: 7⅜.
Step 1. Divide the whole number by 2 and drop the remainder: 7 ÷ 2 = 3.
Step 2. Add the numerator and denominator of the fraction and record that number
         as the half-fraction numerator: 3 + 8 = 11
Step 3. Double the denominator: 8 × 2 = 16.
The answer is 3¹¹⁄₁₆.

Example 4: Halving common fractions and mixed number common fractions.

¼ = ⅛	⁷⁄₁₆ = ⁷⁄₃₂	4⅝ = 2⁵⁄₁₆	7¼ = 3⅝
3⅛ = 1⁹⁄₁₆	5½ = 2¾	6¾ = 3⅜	2⅞ = 1⁷⁄₁₆
⅝ = ⁵⁄₁₆	1¼ = ⅝	8⁹⁄₁₆ = 4⁹⁄₃₂	11³⁄₁₆ = 5¹⁹⁄₃₂

In the previous four units, ¼ of the questions involved the addition of common fractions, ¼ involved the subtraction of common fractions, ¼ involved the multiplication of common fractions, and ¼ involved the division of common fractions. Of course, the addition of the four ¼s equals the whole number of one (1). In this final unit of Section 2, the common fraction problems are mixed so that it will be necessary to perform any one or a combination of the basic mathematical functions to arrive at the correct answer for each of the fifteen problems.

PRACTICAL PROBLEMS

Work the formulas to arrive at the correct answers for the first five problems. For common fraction answers, reduce to the smallest denominator.

1. (¼ + ¼) × 156 =   _____

2. (354 ÷ ⅔) + 34 − 15 =   _____

3. ³⁄₁₆ + ⅛ + ⁷⁄₁₆ − ½ =   _____

4. (15 ÷ ⅓) × 3 − 7 + 38 =   _____

5. ¼ + ³⁄₁₆ + ⅝ ÷ 2 =   _____

6. A customer service representative has to talk with three more customers after she finishes with the current customer. She has been with the current customer for ¾ of an hour and has about another ¼ hour to complete the discussion. She started with the first customer at 8:50 A.M., and she has a luncheon meeting beginning at 12:00 noon. How many minutes can she devote to each of the three remaining customers?   _____

7. A glass-etching specialist had the following amounts of drinking glasses to image before the end of the working day: 1½ dozen 8-ounce drinking glasses, 3¼ dozen coffee mugs, and ⅓ dozen 10-ounce drinking glasses. There were only 5¼ hours left in his working day because of another commitment in the evening. How many glasses and coffee mugs did he have to finish per hour so all three jobs could be finished on time?   _____

8. There were 5½ cartons of 20-pound bond paper in the storage room. Ten reams of bond paper are generally packaged in each carton. The supervisor of the copy center obtained 3⅖ cartons of paper to supply the four high-speed electrostatic copiers. How many reams of 20-pound bond paper remained in the storage room?

   _____

9. The design and layout artist is capable of designing 2⅓ business cards per hour. His usual workday consists of 3½ hours spent designing, and then he doubles as a sales representative for the remainder of his 8-hour workday. How many business cards can he design in a five-day workweek?

   _____

10. In problem 9, the design and layout artist, Daniel, doubles as a sales representative. In this phase of his workday, he finds that it takes an average of ¾ hour to talk with a customer at each face-to-face meeting. On average, it takes ½ hour travel time between customers. On average, how many customers can Daniel see in a four-week period?

   _____

11. In the Common Fractions section of this textbook, there are five units of problems involving the mathematical functions of addition, subtraction, multiplication, and division. Assuming that you have completed all of the problems in Units 7, 8, 9, 10, and up through problem 10 of this unit, Unit 11, what fractional amount (to the lowest fraction) of the problems have you completed?

   _____

12. The instructional units in this textbook are grouped in sections. Some sections include more units than others because of the content. On average, how many units are there in each section?

   _____

13. Sally found that there were 16⅜ reams of paper in her storage cabinet. She needed to use 5⅝ reams for a job that had to be printed and folded within a two-hour time period. She also had another job to complete that required 1⅞ reams of paper. How much paper was left in the storage cabinet after the two jobs were finished?

   _____

14. A sign maker had orders for five signs that would be used in a shopping center, thus they had to be large enough to be seen from a distance. The signs were all the same height, but they had to be of the following widths: 30½", 25¾", 48¾", 16¼", and 58¼" (Figure 11–1). What was the average width of the five signs?

_____

Figure 11–1   Signs of many sizes and shapes are important products in the graphic communications industry.

15. The sign maker sold a second job to the same customer who purchased the five signs identified in problem 14. For this second job, each of the five signs was to be 5⅝ inches shorter because the signs had to be used in special locations where space was limited. What was the average width of the five signs for this second order?

_____

Decimal Fractions

Working with *decimal fractions* is often easier than working with common fractions as presented and experienced in Section 2 of this book. There are, though, some basic rules to remember and use when working with decimal fractions.

A decimal fraction such as 5.284 is a whole number plus one or more numbers following a *decimal* or *dot*. The whole number 5 represents five full units of something while the numbers 284 following the decimal represent a portion or fraction of the next whole number, which, of course, would be 6. The first number following the decimal, in this example 2, represents the *tenths position*. If it were used alone following the decimal, 5.2, the decimal fraction would be read as 5 and two-tenths of something, such as inches or feet. This number could easily be converted to a fraction if necessary: 5.2 equals 5²⁄₁₀, which can be reduced to 5⅕.

With the 5.284 example, the second digit following the decimal, 8, represents the *hundredths position*. If the number was listed as 5.28, the decimal fraction would be read as 5 and 28 one-hundredths, which could also be written as a fraction, 5²⁸⁄₁₀₀, and reduced to 5⁷⁄₂₅. The third number, 4, represents the *thousandths position*; thus, the decimal fraction would be read as 5 and 284 thousandths. If needed, this number could also be listed as a fraction, 5²⁸⁴⁄₁₀₀₀, and reduced to 5⁷¹⁄₂₅₀. The fourth number following a decimal is the 10,000 or *ten-thousandths position*, and the fifth number following a decimal is the 100,000 or *one-hundred-thousandths position*.

Rounding to decimal places is often utilized to save time and space. The basic rule is to round a given number upward if the number or numbers following it are ½ or larger. For example, 2.59 can be rounded to 2.6, and 4.746 can be rounded to 4.75 without significantly changing the overall value of the numbers. Often, the guidelines for working with decimal fractions will be given as round at the third (thousandth) position or second (hundredth) position. In this way, every person, such as a business or industry team or a class of students, will be working under the same rules.

A "rounding" situation that sometimes causes discussion and even debate is whether the number 5 should be rounded up or down. For example, if the instructions are to round 3.435 to two places, should it be 3.44 or 3.43? There is no clear answer, so it depends on the person in charge of the statistics or dollars and cents. It is always best to ask the person in charge about which rule is being followed.

 Unit 12   ADDITION OF
DECIMAL FRACTIONS

BASIC PRINCIPLES OF ADDING DECIMAL FRACTIONS

The addition of decimal fractions is accomplished in the same manner as the addition of whole numbers. The only difference, and it is important, is that the decimal points must be aligned when adding decimal fractions in the traditional "hand" method. The following is provided as an example.

Example:
```
      2.4
    529.62
     48.239
 +   34.5
    614.759
```

To cause less confusion, it is wise to add zeros to the decimal numbers so all of the numbers following the decimal point have the same number of digits. In this example, one of the fraction numbers has three numbers (239) or thousandths. By adding zeros to the other three fraction decimals, the numbers would appear in the following manner:

```
      2.400
    529.620
     48.239
 +   34.500
    614.759
```

When using a calculator to add decimal fractions, it is important to remember the decimal point and position it correctly within each number being added. For example, with the four decimal fractions in the previous example, the numbers should be entered as 2 decimal 4, 529 decimal 62, 48 decimal 239, and 34 decimal 5. The computer chip in the calculator has been programmed to correctly position the decimals according to the order in which each is entered; thus, the correct answer of 614 decimal 759 will be the result.

PRACTICAL PROBLEMS

1. An estimator calculated that it would take the following number of hours to complete a calendar for the local insurance agent: design and layout, 3.75 hours; copy preparation, 6.50 hours; photo/electronic conversion, 1.33 hours; image carriers, 1.80 hours; image transfer, 4.60 hours; and finishing and shipping, 3.40 hours. What is the total number of estimated hours to complete the calendar?

   _____

2. Two different screen presses are being used during a three-day period to print the same long-run textile job. The first screen press is used for 6.5 hours on Monday, 7.3 hours on Tuesday, and 4.6 hours on Wednesday. The second screen press is used for 3.7 hours on Monday, 6.8 hours on Tuesday, and 6.9 hours on Wednesday. What is the total number of hours the two presses are used to print the textile job?

   _____

3. To take advantage of the prepaid freight policy of the North Central Paper Company, the purchasing agent of the small town graphic arts company placed a rather large paper order. She ordered 240.666 pounds of 60-pound text paper, 420.250 pounds of 50-pound uncoated book paper, and 400.789 pounds of 20-pound #4 sulfite bond paper. What was the total weight of the shipment of paper ordered by the purchasing agent?

   _____

4. LaVada, a print buyer, requested and received bids from four commercial printing companies on a variety of jobs. The bid from company A was $4,789.38 for some four-color process brochures; the bid from company B was $2,397.45 for statements and envelopes; the bid from company C was $789.29 for direct-mail flyers; and the bid from company D was $1,807.03 for order forms. What was the total amount of the bids for the four different print jobs?

   _____

5. January expenses for the Creative Pad Printing Company consisted of the following amounts: salaries and related taxes, $33,789.04; utilities, $2,515.89; building rent, $8,750.50; insurance, $479.83; repairs, $348.95; and miscellaneous expenses, $74.73. What were the total expenses for the month of January?

   _____

6. A design and layout artist planned the following measurements for the copy preparation computer specialist to follow when preparing the original copy for a campaign leaflet: left margin, .75 inch; text line length, 3.125 inches; photograph halftone image, 2.5 inches, and right margin, .75 inch (Figure 12–1). What will be the total sheet width of this leaflet?

_____

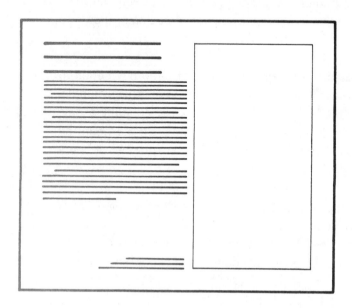

Figure 12–1   This rough layout for a campaign leaflet provides valuable position information for the copy preparation computer specialist.

7. The gross receipts for the *first quarter* at the Quick Turn-Around Imaging Company are listed as follows:

January	$11,289.43
February	$12,769.57
March	$14,863.69

What were the first-quarter gross receipts for the Quick Turn-Around Imaging Company?

_____

8. The operator of a paperfolder recorded the following work hours on his time card for a particular week: Monday, 8.2 hours; Tuesday, 9.5 hours; Wednesday, 8.6 hours; Thursday, 8.2 hours; and Friday, 7.8 hours. How many total hours did the operator of the paperfolder work on the first, second, and fourth days of the week?

_____

9. The litho press operator was required to install a new blanket that included the usual packing. The press operator added two sheets of paper that each measured .057 inch thick, one sheet of paper that measured .040 inch thick, and one sheet of paper that measured .0376 inch thick (Figure 12–2). What was the total thickness of the packing sheets that were placed under the blanket and against the steel cylinder?

_____

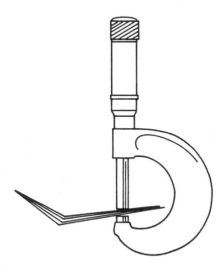

Figure 12–2 The thickness of the litho press blanket packing sheets is measured with a tool called a micrometer.

10. The prepress phase, including the making of litho plates, took considerable time to prepare an advertising piece for a chain of hardware stores. The design and layout work for the job had been completed previously, thus it was available for production to begin. The computer composition work took 4.2 hours, full-color proofs took 1.7 hours, film output from the imagesetter took 1.2 hours, and litho platemaking took .75 hour. How many total hours were charged to the job?

_____

11. A flexography press operator divided her workday as follows: makeready, .75 hour; running time, 4.35 hours; cleanup, .50 hour; and preventative maintenance, 1.9 hours. What were the total hours that the flexography press operator recorded for the day?

_____

12. The purchasing agent of the Quality Graphics Company inventoried the litho ink in the storeroom. The agent found the following types, colors, and amounts of ink: litho-process yellow, 12.4 pounds; litho-magenta, 9.6 pounds; litho-cyan, 7.5 pounds; flexo-cyan, 8.9 pounds; litho-purple, 1.7 pounds; litho-process black, 27.3 pounds; flexo-brown, 4.8 pounds; litho-green, 5.2 pounds; and litho-mixing white, 6.4 pounds. How many pounds of litho ink did the purchasing agent find in the storeroom?   _____

13. The owner of the Smith Graphic Design Agency recently noted that the following accounts are due from his customers by the tenth of next month: account A, $786.54; account B, $1,245.89; account C, $87.95; and account D, $256.74. What is the total of the four accounts due to the Smith Graphic Design Agency?   _____

14. The two estimators from the Quality Graphics Company prepared several estimates during a recent busy Monday. First, they estimated a pamphlet job for $1,278.64; second, they estimated some letterheads for $89.04; third was the estimate for the accompanying envelopes for $174.56; next came the estimate for some three-part carbonless forms for $1,765.89; and finally, they prepared an estimate for some special four-color envelope stuffers for $10,876.93. With the exception of the envelopes because they were canceled, how many dollars of estimates did the two estimators prepare?   _____

15. Four jobs will be ganged on a litho press run. The first job measures 4.25 inches wide, the second job measures 5.5 inches wide, the third job measures 6.75 inches wide, and the fourth job measures 8.5 inches wide. The total trim for the four jobs is 2.3 inches which includes 1.0 inch on the left side and 1.3 inches on the right side. What is the minimum width of paper on which these jobs can be run?   _____

 ## Unit 13 SUBTRACTION OF
DECIMAL FRACTIONS

BASIC PRINCIPLES OF SUBTRACTING DECIMAL FRACTIONS

Subtracting decimals fractions is much like subtracting whole numbers. The major difference is that it is necessary to retain alignment of the decimal points. Of course, this is utilized in the addition of decimal fractions; thus, there is no difference in this procedure. As with adding decimal fractions, it is helpful to add zeros to the numbers following the decimal so all decimal fractions have the same number of digits. The following examples should be of help when working the practical problems.

Example 1: Subtract 25.32 from 73.56.

$$
\begin{array}{r}
73.56 \\
-\,25.32 \\
\hline
48.24
\end{array}
$$

Example 2: Subtract 30.27 from 37.15.

$$
\begin{array}{r}
37.25 \\
-\,30.27 \\
\hline
6.98
\end{array}
$$

Example 3: Subtract 55.238 from 75.4.

$$
\begin{array}{r}
75.400 \\
-\,55.238 \\
\hline
20.162
\end{array}
$$

PRACTICAL PROBLEMS

1. The monthly gross receipts for the ABC Copy Creation House amounted to $20,834.34. Unfortunately, the expenses amounted to $18,750.91, but this did give the owner of the company some profit. What was the profit for the month? _____

2. The weekly financial inventory of paper stock revealed that there was $4,764.29 worth of paper in the storeroom. Thirty days later, a second inventory revealed a drop to $3,975.39 worth of paper in the storeroom. How much less is the second inventory dollar amount than the earlier inventory figure? _____

3. Sales figures are often an accurate barometer of how well a graphic arts-related company is doing. The total sales for the company last September were $24,786.93, whereas the sales for the most recent September were only $21,956.47 (Figure 13–1). By what amount were the sales down this September?

_____

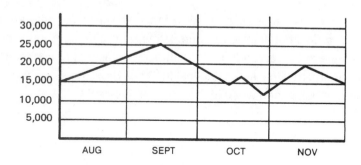

Figure 13–1 Graphs can be used as excellent communicators of information such as sales figures.

4. Real profit is not always shown in the dollar amount of sales, but in how efficiently the jobs are produced. Last year, a four-part NCR form was produced on an older litho press in 21.75 hours. This year, the same job was run as a repeat order and completed in 18.25 hours. Much of the reduced time could be attributed to the acquisition of a new litho press that had a higher operating speed. How much time was saved by using the new piece of equipment?

_____

5. The estimating team of a large litho company determined that the cost of producing a brochure would be $1,345.63. When the job was completed and the figures were tabulated, the actual cost was $1,298.97. For how much under the estimated amount was the job produced?

_____

6. The embroidery department supervisor scheduled a sweatshirt run for 16.7 hours. Because of some thread feeding problems, the job actually took the operators 19.3 hours to complete. How much longer did the sweatshirt run take to complete than was scheduled?

_____

7. A bread wrapper being printed on a two-color flexography press was estimated to use 9.5 pounds of black ink and 3.7 pounds of red ink. After the run was completed, the press operator determined that 10.1 pounds of black ink and 4.2 pounds of red ink were used. How much more ink was actually used than was estimated?

_____

8. An experienced computer operator, using a standard software package, was able to create an eight-page color brochure in 3.6 hours. Working on a similar job, a new employee, straight from a community college graphic communications program, took 6.8 hours to complete the job. How much faster did the experienced computer operator complete the eight-page color brochure than the new employee?

_____

9. Two litho press operators work for the same company. Press operator A, who frequently works overtime, took home $45,585.75 last year. Press operator B, who cannot work overtime because of some other responsibilities, took home $42,360.89 during the same time period (Figure 13–2). What was the difference between the take-home pay of the two press operators?

_____

Figure 13–2   Time-card machines are valuable for recording employee work hours.

10. When the Snodley Communications Company first started business many years ago, the employees in the photoconversion department processed film by the tray method. Several years ago, an automated film processor was purchased that allowed the work to be completed much faster. For 100 sheets of 10" x 12" film, it took 4.69 hours to tray process the film, dry-to-dry. With the automated film processor, it took the operators only 1.90 hours to process the same amount of film. How much time was saved by using the automated equipment over tray processing?

_____

11. A job that was bid for running on a 19" x 25" litho press had a pressroom cost of $57. Because of a breakdown on this press, the job was run on a smaller press which then had a pressroom cost of $153.99. How far over the budgeted amount was the final pressroom cost?

———————————

12. Inventory systems are usually established with a reorder point for every item that is commonly used in production. For the Prairie Color Graphics Company, the normal inventory of black gravure ink was established at 90 pounds. Because of a breakdown in communications among the several employees, the inventory dropped to 19.67 pounds. How many pounds of ink had to be ordered to bring the inventory back to the normal minimum?

———————————

13. Mixing photographic chemicals is routine for employees in a photoconversion department where they have one or more imagesetters. For some rapid-access film, the formula for the working solution of fixer was .50 gallon of water, .25 gallon of solution A, and .05 gallon of solution B (Figure 13–3). In order to complete the mixture, it was necessary to add how much more water to bring the mixture to one full gallon?

———————————

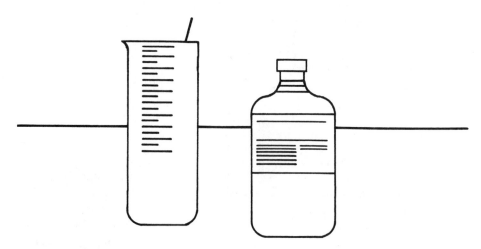

Figure 13–3   Chemical mixing must be done safely and accurately for quality results.

14. To boost sales, the sales manager of the Foothills Prepress Company established a promotional program for her sales representatives. The prize was a weekend trip for two people to a popular resort high in the mountains. Sales representative Jones had sales that amounted to $147,986.47 during the promotional period. During the same time frame, sales representative Nelsen sold $152,797.84 worth of printed products. By what margin did sales representative Nelsen win the trip?

_____

15. A design and layout artist prepared a layout for a job that was about to be created by the copy preparation computer operator. The three halftones took 2.25 inches of horizontal space, the space between the halftones and the paragraphs was .25 inch, and the four paragraphs of copy took 4.625 inches of horizontal space (Figure 13–4). The sheet on which this content will be printed will be 8.5 inches wide. How much total white space will remain for the left and right margins?

_____

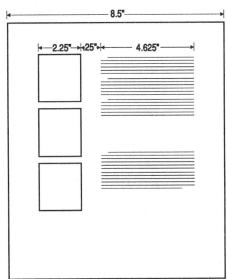

Figure 13–4 The comprehensive layout provides an excellent visual image of copy element locations and measurements.

Unit 14   MULTIPLICATION OF DECIMAL FRACTIONS

BASIC PRINCIPLES OF MULTIPLYING DECIMAL FRACTIONS

The multiplication of decimal fractions is accomplished in the same manner as the multiplication of whole numbers except for one important thing. The product (answer) when multiplying two decimal fractions must include the total of the decimal places in both decimal fractions.

Example: 6.37 x 4.2 = 26.754

There are two decimal places in one number and one decimal place in the other number for a total of three decimal places. As when working with other decimal fractions, it may be permissible to round the decimal digits to one or two places, such as 26.8 or 26.75.

When using a calculator to multiply decimal fractions, it is important to place the decimal in the correct location with each decimal fraction number. Using the same numbers as above, enter (6 decimal 37 x 4 decimal 2 =). The computer chip in the calculator has been programmed to handle the decimals in the correct manner; thus, the decimal will be located in the correct place in the answer.

PRACTICAL PROBLEMS

1. A graphic designer, working for a mirror manufacturer where beautiful images are etched on mirrors of all sizes and shapes, earns $11.75 per hour and works an average 37.5-hour week. What is the gross weekly pay for the graphic designer?   _____

2. Copy preparation department personnel spent 3.5 hours preparing a small job that had to be ready by 4:30 P.M. on Thursday. The estimator charges out the department work at $38.75 per hour. What was the total cost to the customer for this job?   _____

3. After the monthly inventory, it was found that there were 47.33 reams of 20-pound bond paper on the shelves. Upon reviewing the pricing catalog, the inventory supervisor determined that this paper was priced at $65.89 per ream. What was the total value of the 20-pound bond paper currently in inventory?   _____

4. A trade business card company located in the northwest United States only sold their products through dealers distributed throughout the country. On Monday of a given week, 497 orders were sold and delivered, and on Tuesday of the same week, 513 orders were sold and delivered. The average selling price per order of these cards to the dealers was $9.48. What was the total gross sales of business cards for the two days?   _____

5. A catalog that is being produced by the skilled employees at the Rivers Edge Publishing Company will require 75 10" x 12" pieces of film that can be produced on the imagesetter. A typical roll of film that is purchased for the imagesetter yields 100 sheets of 10" x 12" film and costs $87.98 per roll. What will be the cost of the film used for the 75-page catalog?   _____

6. In the litho press department of the Barton Specialty Company, 3.7 hours are required to print a two-color letterhead job. The standard operating cost of the press that is used is $20.65 per hour. What is the charge for the press time needed to print the two-color letterhead?   _____

7. There are six columns in a tabular that are each 1.67 inches wide (Figure 14–1). The margins or space between each column will be ¼", and the left and right margins will be .75 inch each. What is the total width of the sheet that will be needed for this six-column job?   _____

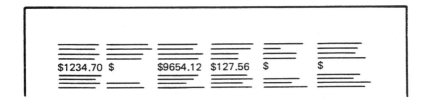

Figure 14–1   A rough layout of a tabular form containing six columns.

8. The two employees in the binding and finishing department earn an average wage of $9.69 per hour, and the regular workweek is 37.5 hours. Employees are paid time and a half for any overtime hours, and this past week, one of the employees worked 43.25 hours. What was the gross wage of the employee who worked the extra 5.75 hours?   _____

9. A copy preparation computer operator is capable of keyboarding 25.5 words per minute. On Friday, the operator worked 1.875 hours keyboarding the Sunday morning worship service bulletin for a local church (Figure 14–2). Based on the computer operator's production rate, how many words were included in the church bulletin?

_____

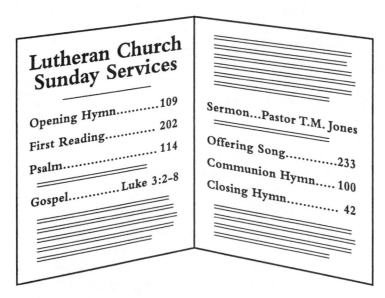

Figure 14–2 Church bulletins are major income products for many graphic communications companies.

10. The average monthly sales for the 24-hour Images-and-More Company for the previous year were $19,276.43. Based on this average, what were the total annual sales?

_____

11. For a book cover that contained a large halftone image and some heavy line copy, it was taking 5.66 pounds of black ink per hour to print 25,000 copies with a medium-sized litho press. To complete this press run, it took 3.75 hours not including makeready and cleanup. How much ink was used to print the 25,000 copies of the book cover?

_____

12. Dawn's Graphic Services Company is an expanding company with three full-time sales representatives. For the first quarter (13 weeks) of the year, the average sales of their 5.5-day workweek for the three sales representatives amounted to $4,674.89. What was the total amount of sales for the first quarter of the year?

_____

13. The graphic arts company estimator has provided a sales representative with figures for a brochure job. The job is estimated to take 3.6 hours to fold, and the paperfolder hourly cost rate is $37.50. What will be the cost of folding the printed brochure? _____

14. The inside of an 108-page book (printed one side) was printed on paper that measures .0035 inch per sheet (Figure 14–3). What is the total thickness of the body of the book? _____

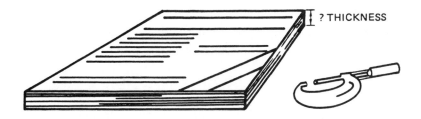
? THICKNESS

Figure 14–3 Micrometers are used to measure the thickness of individual sheets of paper so the total book thickness can then be determined by simple multiplication.

15. A photo/electronic conversion operator, who is paid $9.75 per hour, works 2.3 hours on an advertising poster. The screen printing stencil maker, who is paid $10.25 per hour, works 1.7 hours on the same job. How much labor cost is charged to the advertising poster based on work from these two people? _____

Unit 15   DIVISION OF DECIMAL FRACTIONS

BASIC PRINCIPLES OF DIVIDING DECIMAL FRACTIONS

The division of decimal fractions is accomplished in the same manner as the division of whole numbers with one exception. That exception is, of course, the decimal points. If the divisor is a decimal fraction, it must be converted to a whole number by moving the decimal point to the right and actually eliminating it. The decimal in the *dividend* number must then be moved to the right the same number of digits as in the divisor whether it is a decimal fraction or a whole number. After the decimal points have been moved, the division can be done in the same way as with whole numbers. The following examples should be of help.

Example 1:   Divide 149.35 by 20.60.

```
                                          7.25
  20.60 )149.35 ---------------------  2060 )14935.00
                                            14420
                                             515 0
                                             412 0
                                             103 00
                                             103 00
```

Example 2:   Divide 125 by 8.75.(Hint: Round at two places and add zeros as needed.)

```
                                        14.2857  = 14.29
  8.75 )125 -------------------------  875 )12500.0000
                                            875
                                            3750
                                            3500
                                             250 0
                                             175 0
                                              75 00
                                              70 00
                                               5 000
                                               4 375
                                                 6250
                                                 6125
                                                  125   (past the two places of rounding)
```

Example 3: Work this problem with a calculator. A press operator placed 2.6 pounds of ink in the ink fountain of the litho press he was operating. He proceeded to print 6,550 sheets of stock before it was necessary to add ink. How many sheets of paper were printed per pound of ink?

$$6,550 \div 2.6 = 2519.230769$$

Round at zero place because it is impossible to print a partial sheet of paper. Thus, 2,519 sheets were printed with each pound of ink.

PRACTICAL PROBLEMS

1. During an average week of 4.5 days, a sales representative for a prepress company had total sales that amounted to $6,500.16. What was the average sales per day? _____

2. On a recent workday consisting of 7.5 hours, a litho press operator printed 33,750 #10 envelopes. What was the press operator's average production rate? _____

3. According to the graphic arts company estimator, a 28,600 single-color label can be printed on the flexography press in 3.25 hours. How many impressions per hour must the press operator maintain to meet the estimate? _____

4. At the Aaron Communications Company, the purchasing agent ordered 3,000 reams of #4 sulfite light green bond paper. It was to be cut 8½" x 11" and delivered by the Heartland Paper Company. The total cost of the order, including the cutting and delivery, was $13,125. Based on this information, what was the cost of each ream of paper? _____

5. In a monthly newsletter, the average line of type contained 68.75 characters. For this particular kind of type, 2.5 characters could be set per pica. How many picas long was each line of type? _____

6. The image carrier supervisor recommended that 350 long-run plates be purchased for the 19" x 25" litho press. The company purchasing agent informed the supervisor that the total plate cost would be $1,522.50. What was the cost of each litho plate? _____

7. The total bill for a batch of company letterheads, envelopes, and business cards was $650.45. The sales representative noted that this amount was higher than the original estimate, so she requested a breakdown of the bill. It was noted that the jobs required 16 pieces of film for a cost of $59.28 and eight litho plates costing $41.65. What was the cost for each piece of film?   _____

8. In a further investigation of the print job explained in problem 7, the sales representative discovered that the total press time cost the customer $98. Besides the actual cost, it took the press operator 2 hours and 40 minutes to complete the press run. What was the cost per hour for the press time?   _____

9. Cindy, a sales representative, submitted a job to the estimating department for the purpose of obtaining the cost of a job that she was about to sell to a long-time customer. Besides the costs associated with the prepress portion of the job, the press cost was estimated at $132.09 for 3.7 hours of running time. What was the cost per hour for this particular press?   _____

10. The Uptown Design Agency charged a printing company $175.80 for artwork that took 1 hour and 40 minutes. Based on this information, it can be assumed that the hourly rate for this agency is how much?   _____

11. An inventory of the paper stock left in the storeroom reveals that there is a total of 29.25 reams of book paper on hand. The inventory supervisor has reported that the book paper stock is valued at $902.28. From this information, what is the cost of the paper per ream?   _____

12. To give estimates on various print jobs, the personnel in the estimating department must know unit prices on such items as rolls of film for the imagesetter and computer-to-litho plates. Each roll of film costs $111.29, and 104 sheets of 10" x 12" film can be acquired from each roll. Computer-to-litho plate material costs $152.95 per roll, but only 83 plates can be obtained from a roll of material because the plates are longer than the film. What is the cost for each individual litho plate?   _____

13. A 17.5-inch sheet of plastic material will be printed and cut into ten equal pieces for use as promotional bookmarkers (Figure 15–1). What will be the width of each piece of bookmark material? _____

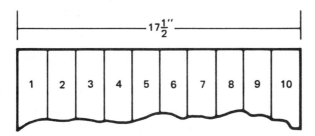

Figure 15–1 A rough layout of a plastic sheet that will be cut into ten equal pieces.

14. A number of 36-pica lines of type will be keyboarded by the copy preparation computer operator (Figure 15–2). The type style and size is Century medium, 14 point. It will be possible to include 67 typewriter style letters into each line. To the nearest thousandth, how many characters per pica are contained in each line of type? _____

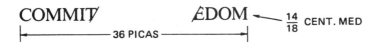

Figure 15–2 If 67 typewriter style characters can be included in a 36-pica line length, how many characters are in each pica?

15. In checking the amount of paper stock on hand, a press operator discovered that 10.4 reams of paper were available. (Remember, a ream equals 500 sheets.) The booklet job will consume four sheets each. How many four-page booklets can be printed from the available paper? _____

 **Unit 16   COMBINING COMMON FRACTIONS AND DECIMAL FRACTIONS**

BASIC PRINCIPLES OF COMBINING COMMON FRACTIONS AND DECIMAL FRACTIONS

When there is need to perform mathematical functions with a combination of common fractions and decimal fractions, it is necessary to *convert* decimal fractions to common fractions or vice versa. A commonality is needed in the type of fractions so the mathematical functions can be completed. It would be difficult to impossible to work a problem such as the one in the following example without conversion.

Example: Multiply ¾ by .258 by ⅜ by 61. A convenient method for performing this multiplication would be to change the common fractions into decimal fractions and then make the calculation. This is done in the following manner:

    a. Convert ¾ to a decimal fraction by dividing the numerator by the denominator:
        $3 \div 4 = .75$.

    b. Convert ⅜ to a decimal fraction by dividing the numerator by the denominator:
        $3 \div 8 = .375$.

    c. $.75 \times .258 \times .375 \times .61 = 0.044263125$, rounded to three places $= 0.044$.

    d. If needed, convert the decimal fraction to a common fraction, $^{044}/_{1000}$, and reduce to $^{22}/_{500}$, then to $^{11}/_{250}$.

PRACTICAL PROBLEMS

1. The head estimator of Seth Legal Publications, Inc., estimated the amount of paper that will be needed to produce a quantity of legal abstracts for a local law firm. The abstracts will take 10¾ reams of legal paper, 8½" x 14", and 5.5 reams of 8½" x 11" paper. How many reams of paper, disregarding the two different sizes, will be needed for the abstracts?   _____

2. At the end of September, the litho press operator conducted the monthly inventory of the ink supply (Figure 16–1). On one shelf, he found 15.25 pounds of soy-based black ink, 8.75 pounds of soy-based emerald green ink, and 9.5 pounds of petroleum-based purple ink. Upon inspecting the ink, he found that ⅓ of the ink was old and had to be discarded according to proper disposal procedures. How many pounds of ink were discarded?   _____

Figure 16–1   The monthly inventory of the ink supply is an important job of a litho press operator.

3. A binding and finishing company operated by three brothers specialized in books for children. For a recent contract, they completed and shipped 30½ cases of farm animal books, 44.5 cases of circus books, 16.75 cases of fire fighter books, and 27¼ cases of police officer books. How many total cases of children's books were completed and shipped by the employees of the binding and finishing company?   _____

4. The owner of a screen printing company, specializing in college and university sports apparel, purchased 76.75 cases of unprinted sweatshirts. The owner then requested that his employees print 15½ cases of the sweatshirts immediately for one of their good customers. How many cases of sweatshirts were still available for other jobs that would soon be sold?   _____

5. The purchasing agent of Ann's Law Office ordered one and one-half dozen printed ball-point pens for each of the 35.66 full-time equivalent lawyers in the law firm. How many dozen ball-point pens were ordered?   _____

6. An imagesetter operator determined that 15⅔ rolls of film had been exposed and processed during the second week of a recent month. The typical workweek in this graphic arts company was 5.5 days. How many rolls of film were used each half-day of the 5.5-day week?

    _____

7. The accountant of a flexography-based packaging company determined that 549.75 cartons of printed paper sacks had been shipped in a one-week period. Unfortunately, 23¼ cartons were returned because of some imperfections. The accountant could invoice the customer for how many good paper sacks?

    _____

8. The ink-making specialists at Tone Pack Inks were preparing the ingredients for an order of 67.25 pounds of soy-based purple ink (Figure 16–2). They needed to add ¾ pound of vehicle for each pound of finished ink. How many pounds of vehicle did the ink-making specialists add to this batch of purple ink?

    _____

Figure 16–2   Calculating the correct amount of an ingredient in a batch of printing ink is essential for the ink makers. (Credit: Flint Ink Corporation.)

9. The print buyer of a large department store was asked by the store manager to purchase a supply of envelopes for the next series of promotional mailings. After some research, the print buyer ordered the following sizes and amounts of envelopes in boxes of 500 envelopes each: 5.5 boxes of #10 white, business; 8¼ boxes of #10 white, window business; 6½ boxes of #10 light blue, business; and 12.25 boxes of #6¾ white, window for invoicing. What was the total number of individual envelopes ordered by the print buyer?

    _____

10. After the original envelope order described in problem 9 was placed, the print buyer decided that it would be wise to order three times as many #10 light blue, business envelopes and reduce the order for the #6¾ white, window envelopes by 4.75 boxes. How many individual envelopes of these two sizes, colors, and styles were finally ordered by the print buyer?

    _____

11. A recent associate of arts graduate of a community college graphic communications program was serving as a supervisor of a five-employee prepress department. The supervisor was charged with completing a large job that would take a total of 12.5 additional hours of work for all five employees beyond the standard 37½-hour workweek. How many hours of overtime would be necessary for each of the five employees to finish the job according to the estimated additional hours?

    _____

12. The prepress supervisor described in problem 11 found that the job could not be finished in the estimated 12.5 hours of overtime for the five employees. In fact, he determined that an additional 6.75 hours would be needed. What was the grand total of hours worked by the five employees during the standard 37½-hour workweek plus the overtime hours?

    _____

13. The standard production time for 100 copies of a 48-page booklet was determined for the following four pieces of finishing and binding equipment: paperfolder = .75 hour; wire stitcher = ⅔ hour; paper cutter = .33 hour; and paper drill = ¼ hour (Figure 16–3). What was the standard production time to complete the 100 copies of the 48-page booklet for all four pieces of finishing and binding equipment?

_____

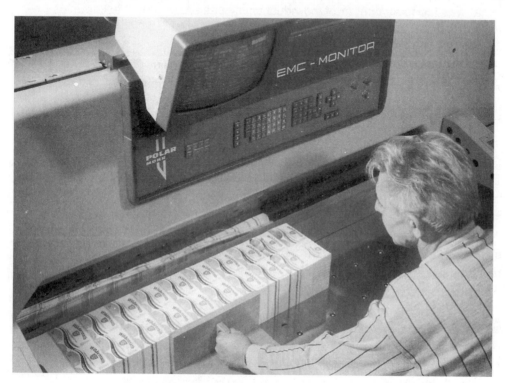

Figure 16–3 For correct estimating, both for cost and time to produce a product, the standard production time must be accurately determined. (Credit: Heidelberg USA, Inc.)

14. A litho press operator was responsible for running three jobs during a standard 7½-hour workday. One job took 3.4 hours, and the second job took 2.2 hours. How many hours were left to complete the third job?

_____

15. Neal, a graphic arts products sales representative, called on six customers during a recent successful day. He sold 6.5 cartons of litho wipes to customer A and 9.25 cartons of litho wipes to customer B. Customers C, D, and E each purchased an even 12 cartons each. Upon arriving home, Neal found two messages on his telephone voicemail. Customer A had decided to increase her order by 2½ cartons, but customer B had discovered a supply of litho wipes in the storeroom and had decided to reduce his order by 6¾ cartons. Based on the five original orders and the two revisions, what was the total number of cartons of litho wipes that Neal sold that day?   _____

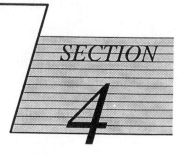

Percent

The mathematical principle of *percent* is very useful when making a variety of calculations. Percent can be defined as a portion of the whole. For example, say a student earned a 90% grade on the midterm examination. If there were 100 questions on the midterm examination, then the student got 90 of them correct. A percent or *percentage* can be calculated on any number including the number one (1). If there was only one mathematical problem to complete, it would be possible to get 25%, 50%, or any other percentage of it correct including 100%.

Percents can be listed with the percent symbol "%" or as a decimal fraction such as .84. This means that .84 is or could be 84% of something. Using a course final examination as an example, say that there were 150 questions or points on the examination. The teacher informed Sally that she got 84% of the 150 questions correct; thus, Sally got 126 questions correct.

An important thing to remember when working with percents is that the "%" symbol implies that the number it is being used with contains two decimal places, as with the 84% or .84. Another example might be a number such as 25.19%. If this number were to be used to multiply times another, the number would be written as .2519 and the % symbol would be left off.

Changing a percentage to a fraction can be accomplished very easily. For example, 84% = $^{84}/_{100}$, reduced to $^{42}/_{50}$ to $^{21}/_{25}$. A percentage number, unless it already contains decimal places, is a percent of 100, which is why 84 is placed over 100 in the example. When calculating a percentage, the formula is:

$$\text{Part } (P) = \text{Percent } (\%) \times \text{Whole } (W)$$

or

$$\text{Percent } (\%) \times \text{Whole } (W) = \text{Part } (P)$$

or

$$\text{Whole } (W) = \text{Part } (P) \div \text{Percent } (\%)$$

The following examples should be of some help.

Example 1: Find 30% of 125 feet.

.30 × 125 = 37.5 feet

Example 2: What percent of 45 feet is 12 feet?

% = P/W

% = $^{12}/_{45}$ = 0.26666666 = 26.67%

Example 3: The number 22 is 40% of what number?

$W = P/\%$

$W = 22/.40$

$W = 55.0$

Unit 17   PERCENT EQUIVALENTS

BASIC PRINCIPLES OF PERCENT EQUIVALENTS

Expressing Percents as Decimals

Percent means per hundred; thus, to express a percent as a decimal, it is necessary to divide by 100. To divide by 100, it is a simple step to move the decimal two places to the left of its original location.

Examples: 1. 3% = .03; both numbers mean the same.

2. 63½% = 63.5% = .635; all three numbers mean the same.

Expressing Percents as Fractions

To express a percent as a fraction, it is necessary to divide by 100. The division should be done as a fractional problem.

Examples: 1. $3\% = 3 \div 100 = \frac{3}{1} \times \frac{1}{100} = \frac{3}{100}$

2. $66\frac{2}{3}\% = 66\frac{2}{3} \div 100 = \frac{200}{3} \times \frac{1}{100} = \frac{200}{300} = \frac{2}{3}$

Expressing Decimals as Percents

To express a decimal as a percent, it is necessary to multiply by 100. To multiply by 100, it is a simple step to move the decimal two places to the right of its original location.

Examples: 1. .75 = 75%

2. .387 = 38.7%

Expressing Fractions as Percents

To express a fraction as a percent, it is necessary to multiply by 100. The multiplication procedure should be done as a fractional problem.

Examples: 1. $\frac{3}{5} = \frac{3}{5} \times \frac{100}{1} = \frac{300}{5} = 60\%$

2. $\frac{5}{6} = \frac{5}{6} \times \frac{100}{1} = \frac{500}{6} = 83\frac{1}{3}\%$

PRACTICAL PROBLEMS

A. Express the following percents as decimals (Figure 17–1).

1. 15% _____   6. 83%   _____

2. 30% _____   7. 925%   _____

3. 145% _____   8. 3¼%   _____

4. ½%   _____   9. 4.81% _____

5. 8½% _____   10. ⅞%   _____

$1.00	one dollar
$10.00	ten dollars
$100.00	one hundred dollars
$1000.00	one thousand dollars

Figure 17–1   Knowing where to place the decimal can make the difference in a profit or a loss situation.

B. Express the following percents as fractions.

11. 30%   _____   16. 16⅔%   _____

12. 60%   _____   17. 33⅓%   _____

13. ¾%   _____   18. 475%   _____

14. 56%   _____   19. 12.5% _____

15. 6¼%   _____   20. 87.5% _____

C. Express the following decimals as percents.

21. .2   _____   26. .00875 _____

22. .45   _____   27. .65125 _____

23. .315 _____   28. 2.667   _____

24. .0037 _____   29. .001   _____

25. .625 _____   30. 1.02¾   _____

D. Express the following fractions as percents.

31. $\frac{2}{5}$ _____   32. $\frac{7}{50}$ _____

33. $1\frac{1}{4}$ _____   34. $\frac{13}{20}$ _____

35. $\frac{3}{10}$ _____   36. $\frac{1}{8}$ _____

37. $1\frac{5}{8}$ _____   38. $\frac{18}{50}$ _____

39. $6\frac{2}{5}$ _____   40. $\frac{9}{16}$ _____

Unit 18   SIMPLE PERCENT

BASIC PRINCIPLES OF SIMPLE PERCENT

The formula for solving percent problems is

$$P = B \times R$$

R is the rate or the number with the percent sign next to it (for example, 8%), *B* is the base or the original quantity, and *P* is the percentage or the ending amount. A great help in mathematics is to remember that "of" means *times* (×) and "is" means *equals* (=).

PRACTICAL PROBLEMS

1. 45% of 819 =   _____

2. 12½% of 800 =   _____

3. How much is 9% of 164?   _____

4. ⅝% of 600 =   _____

5. What number is 374% of 647?   _____

6. The estimator calculated the cost of the paper stock for a two-color flyer to be $36.90 and the cost of printing the flyer to be $18.75. The estimator figured the company would make 20% profit on the paper stock and 30% profit on the cost of the actual printing. What is the total amount of the bill for the two-color flyer?   _____

7. The manager of an insurance company in-plant facility purchased 800 sheets of a special cover paper for an annual report of one of the insurance company divisions. Unfortunately, 15% of the paper was damaged by water from a pipeline break. The press operator went ahead and used 75% of the remaining paper to print the covers for the annual report. How many sheets were still left from the original 800 sheets of cover paper?   _____

8. The Cedar Falls Paper Company inventory supervisor noted that they had 7,439 reams of "1847" bond paper in their warehouse on January 1. During the month of January, 23% of that paper stock was sold. The following month, another 37% of the amount of that same paper that was on hand at the first of February was sold. Also, 5,487 reams of this same kind of paper were received from the papermill during February. What is the actual amount of "1847" bond paper on hand on March 1?   _____

9. During the last fiscal year, the gross income of BAD Graphics, Inc., amounted to $831,500. This income was distributed in the following manner: wages 42%, materials 17%, new equipment and repairs 16%, taxes 2%, interest 3%, reserve 6%, profit 9%, and miscellaneous 5%. How much money was allocated for wages and taxes combined?   _____

10. 4 is _____% of 40?   _____

11. 5 is what percent of 8?   _____

12. What percent of 48 is 2?   _____

13. 9.6 is what percent of 6.4?   _____

14. What percent of $171 is $135?   _____

15. An invoice that amounts to $279.65 for letterheads and envelopes is submitted to a customer. In this amount, $55.00 is included as profit. What is the percent of profit for this transaction?   _____

16. The estimated time for press makeready on a 250,000-length run was calculated at 4½ hours. The litho press operator, though, was able to complete the makeready in 3¾ hours. What was the percent of time gained through the efficient and diligent work of the litho press operator?   _____

17. The owner of the ALD Publishing Company purchased a used minivan to serve as a delivery vehicle (Figure 18–1). It was necessary to take a loan for $9,875.00, and of that amount, $839.75 was the total interest paid on the loan. What was the interest rate for the loan? _____

Figure 18–1 Delivery vehicles are expensive; thus, it is important to negotiate the lowest borrowing percentage rate as possible.

18. The hourly pay rate of a bindery worker was $8.20. Recently, this person received an increase of $1.00 per hour to raise his hourly pay to $9.20. What was the percent of the raise for the bindery worker? _____

19. The estimated time for preparing the copy for an advertising flyer was 4.2 hours. Unfortunately, the copy preparation specialist encountered some problems and it took five hours to complete. What was the percent of efficiency for the copy preparation specialist? _____

20. 44 is 55% of what number? _____

21. 5% of what number is 45? _____

22. 513 is 114% of what number? _____

23.   ½% of what number is 10?   _____

24.   28% of what number is 28?   _____

25.   A sales representative of SBD Legal Graphics sold a print job for $1,260. This calculated out to be 12% more than it cost. What did this job cost to produce?   _____

26.   A paper sales merchant sold a quantity of paper at a profit of 12½% and gained $250. What was the original cost of the paper?   _____

27.   A printed job that cost $865 to produce was sold for a 15% profit. Another job that cost $1,210 to produce was sold for an 8% loss. What was the selling price of both jobs combined?   _____

28.   A materials bill for $980.50 was submitted to the binding and finishing company vice president by a graphic arts sales representative. Of that amount, 12% was charged for profit. How much profit did the graphic arts sales representative make?   _____

29.   The owner of a quick printing business sold an order of company letterheads for $1,750. Unfortunately, the job was completed several days after the established deadline, thus the owner of the quick printing business was required to deduct 8% from the original quotation. How much money could the owner expect from this sale?   _____

30.   A recent graduate of a university graphic communications program found employment for $425 per week. The graduate decided that it would be wise to place $50 in a savings account each week. What percent of the graduate's income was being saved?   _____

 # Unit 19   CALCULATING INTEREST

BASIC PRINCIPLES OF CALCULATING INTEREST

Interest is the amount of money paid for the use of borrowed money. Interest is charged as a percent of the amount borrowed. The interest rate is given for one year, often called *per annum*, but the money may be borrowed for more or less than one year. The *principal* is the amount of money borrowed. This formula is typically used for calculating interest:

$$I = P \times R \times T$$

whereas, I = Interest, P = Principal, R = Rate (in either decimal or fraction form), and T = Time (in years). For purposes of this unit of study, 12 months equal one year and 365 days equal one year. The following completed problems should serve as appropriate examples for working the fifteen problems in this unit.

Example: Find the interest on $5,280 at 10½% for three years.

$I = P \times R \times T$

$I = \$5,280 \times .105 \times 3$

$I = \$1,663.20$

Example: Find the interest on $6,475 at 9¾% for eleven months.

$I = P \times R \times T$

$I = \$6,475 \times .0975 \times {}^{11}\!/_{12}$

$I = \$578.70$

PRACTICAL PROBLEMS

1.   An employee of a large book publishing company deposited $1,245 into the company credit union. The employee was informed that the annual interest rate of return was 5½% per annum when the deposit was made. How much interest did the employee receive at the end of the first year?   _____

2. The owner of the small Quick Print Shop, located in the city shopping mall, needed to purchase some equipment costing several thousand dollars. Not having sufficient reserve, the owner had to borrow $7,850 to help cover the cost of the equipment. He agreed to repay the loan in nine months at an interest rate of 15% per annum. What amount of interest did the Quick Print Shop owner have to pay at the end of the nine-month period?

    _____

3. Interest on capital investment should be charged by every owner of a graphic communications company as an item cost of doing business. The Quick Print Shop owner from problem 2 was informed by his accountant that the prepress equipment was valued at $8,040, the press equipment (before the new purchase) was valued at $34,460, and the binding and finishing equipment was valued at $22,420. The accountant stated that the appropriate interest rate to be charged against this equipment should be 16%. How many dollars should be charged against this equipment for one calendar year?

    _____

4. A computer operator in a copy preparation department purchased a used automobile for $7,865. After a $3,000 down payment, a chattel mortgage was established for the remainder of the automobile cost. The buyer agreed to pay off the loan in 18 months and was told that the total interest charge would be $656.78. What was the interest rate for the loan?

    _____

5. A paper salesperson purchased a home for $86,000. In addition to interest, at the rate of 8.3% per annum, the salesperson agreed to reduce the $86,000 mortgage by $500 every six months. What was the amount of interest paid at the end of the first month?

    _____

6. For this problem, the figures and conditions described in problem 5 are to be utilized. What was the total amount of interest due for the second six-month period?

    _____

7. The net income of Jones Graphic Services for this past year amounted to $15,128. The owner of the company usually figured a 17% net return on their capital investment that included the following values: prepress, $27,674; press, $86,375; postpress, $34,385; and all other investments, $26,465. What percent of return was actually received on the capital investment of the Jones Graphic Services company for the past year?

    _____

8. During the fourth quarter of the current year, the owner of the Jones Graphic Services company calculated that the company would receive a 2.3% higher net return on its capital investment by the end of the year than that gained in problem 7. Assuming that the capital investment remained the same, how much did the net income have to total for the current year to make this happen?   _____

9. An employee of a university in-plant imaging service deposited $500 in his savings account on January 2. On July 2, he again deposited $500 in his savings account knowing that the interest rate was 5½% compounded semiannually. With this amount of principal and the interest rate, how much money did the employee have in his savings account at the end of the first year?   _____

10. Interest in bank and credit union savings accounts usually begins the first of each month. A graphic arts customer service representative deposited $50 each month for six months in her savings account. At the end of the initial six-month period, a report was provided indicating that she had earned $8.75 interest on her savings. What *annual* rate of interest does this represent for the period?   _____

11. The supervisor of the binding and finishing department of a commercial printing company recommended the purchase of a new paper cutter that was priced at $15,467. The owner of the company borrowed $12,000 at 10.4% per annum and then repaid the loan after nine months. How many dollars of interest were paid for the nine-month loan?   _____

12. The owner of the commercial printing company described in problem 11, then had to sell the new paper cutter at the end of the nine-month period to cover a previous loan that was coming due. He was fortunate to sell the new paper cutter for $16,225 because he was willing to move and install the paper cutter at the buyer's site. How much money, not considering the moving and installation costs, did the company owner gain or lose (indicate which) when considering the loan interest cost and the resale of the paper cutter?   _____

13. When purchasing stock in a communications technology–related company, Mr. Byron paid $864 each for 12 shares of stock. The market value of the stock was $929 for each share of stock at the time of purchase. Mr. Byron was advised that income tax of 22% had to be paid on the difference between the purchase price and the market value. How much tax did Mr. Byron have to pay on this transaction?   _____

14. The three owners of a corporate sign-making company purchased a new computer, scanner, and full-color printer/cutter so they could make signs for office doors and foyers of corporate office buildings (Figure 19–1). The computer cost $4,789, the scanner cost $2,135, and the printer/cutter cost $6,254. To make a reasonable profit, the owners believed that they needed to gain a net profit of 16.5% per year on their capital investment. What did their net profit have to be in actual dollars to meet this objective?   _____

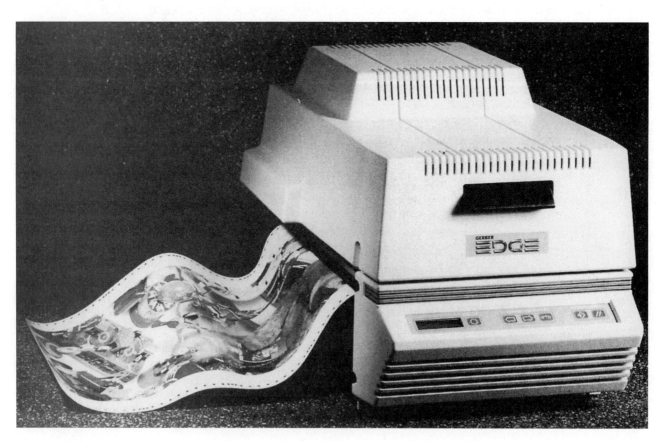

Figure 19–1   A necessary equipment item for operating a corporate sign-making business. (Credit: Gerber Scientific Products)

15. Based on the situation presented in problem 14, the three owners of the corporate sign-making company actually acquired a net profit of $3,728 at the end of their first year of operation. What was the actual percentage of net profit based on their capital investment? _____

Unit 20 CALCULATING DISCOUNTS

BASIC PRINCIPLES OF CALCULATING DISCOUNTS

A *monetary discount* is made on the stated price of a product or service based on a given percent. The percent discounted is determined by several people or someone in charge of product or service sales. This could be the company board of directors, the owner, or the sales manager. The discounted amounts may range from as little as one percent up to 99%. After-holiday sales are frequently based on percent discounts such as "everything in the store will be discounted 50% this coming Saturday and Sunday." Two examples are provided to demonstrate the principles of discounts.

Example 1: Determine the savings on the purchase of $365 of ink if it is discounted 20%.

$ 365.00 (original price) × .20 (discount rate) = $73.00 (discounted savings)

Example 2: Determine the price paid for $365 of ink if successive discounts of 20% and 2% are given.

$365.00 (original price) × .20 (discount rate) = $73.00 (discounted savings)
$365.00 (original price) − $73.00 (1st discount) = $292.00 (discounted price)
$292.00 (discounted price) × .02 (discount rate) = $5.84 (2nd discount)
$292.00 (discounted price) − $5.84 (2nd discount) = $286.16 (price paid)

PRACTICAL PROBLEMS

1. A department store customer noted that all items on a particular table were marked 10% off the price tag amounts. The customer selected a product that was priced at $14.29. How much money was discounted from the original price? _____

2. A customer walked into a postal supplies retail outlet with the intent of purchasing a number of items that would be needed for a mailing to several people. The customer purchased the following items: 250 sheets of preprinted stationery, price marked at $12.65 but discounted 10%; 250 #10 envelopes, price marked at $9.35 but discounted 10%, and 250 calendars, price marked at $70.00 but discounted 15%. What was the total cost of these three items? _____

3. An insurance sales representative had the following items printed at the local commercial printing company: 5,000 letterheads at $9.85 per 1,000; 5,000 envelopes at $7.25 per 1,000; and 2,500 circulars for $34.50. The company payment policy was that if customers paid their bills in 30 days, they would receive a 2% discount. What was the discounted cost of the three items purchased by the insurance sales representative? _____

4. A graphic arts sales representative named Lee was a master at selling imaged products because of the excellent discounts that he was permitted to offer. For a recent sale of $160.80, Lee could offer the customer a 30% discount if one of two standard ink colors was selected. An additional discount of 10% could be given, according to company policy, if payment was made at the time of the order. How much would the customer pay for the order if both conditions were met? _____

5. Lee, the graphic arts sales representative, sold a job to another customer for 2½ times the amount paid by the first customer identified in problem 4. Lee offered the second customer the same discounts he offered the first customer. What was the cost of the work completed for the second customer assuming that both conditions were met? _____

6. The accountant employed by an average-sized graphic communications company worked out a payment plan for one of their good customers. The bill amounted to $3,215.38. If the bill was paid in five working days, a 35% direct discount could be offered; and if the bill was paid in two installments of ten days and then again in ten days, successive discounts of 20% and 15% could be offered. What is the difference in value between the two discount plans for this customer? _____

7. A publisher purchased two tons of paper in roll form for use in printing a monthly magazine. The cost of the paper per ton was usually $334, but it was discounted 10% because it was a discontinued color. The shipping cost was $1.25 per 100 pounds. What was the total cost of the delivered paper? _____

8. A manufacturer of drinking glasses produced 150 dozen drinking glasses with etched logo images for a local baseball team (Figure 20–1). The team owner planned to use the drinking glasses as a promotional for those fans who were the first to buy tickets for the Saturday afternoon game. The calculated cost for each drinking glass amounted to $1.57. With a 5% discount for paying the invoice in ten days, what did the baseball team owner pay for the 150 dozen drinking glasses?

_____

Figure 20–1   Drinking glasses and other glass-based products can be imaged by using an etching procedure that results in excellent detail.

9. Following the first very successful drinking glass promotional, the baseball team owner (see problem 8) decided to order another 150 dozen imaged drinking glasses priced at $1.57 per glass. This time he received a 2.5% discount for using the same image as that used for the first order, and the same 5% discount was offered for paying the invoice in ten days. What was the bottom-line cost for the second order of drinking glasses?

_____

10. The business manager of a business forms printer has been informed by the company estimator that the cost of producing 6,000 business forms will be $450. What will the customer be billed if 20% is allowed for profit?

_____

11. A graphic designer for a newsletter publisher purchased $2,450 worth of home furnishings for her apartment. This amount is the budget price that is payable in 24 monthly installments. At the time of the purchase, she was informed that the cash price would be 12½% less because some of the items were listed as close-out. Being a smart businessperson, the graphic designer negotiated an additional 2% discount based on the fact that she was just getting started in furnishing her apartment and would very likely be purchasing more furniture in the near future. What was the final cost of her first furniture purchase?

_____

12. A quick-print manager purchased 20 pounds of primrose yellow soy-based ink for $189. A direct-purchase discount of 12½% was allowed, and if the bill would be paid in 30 days, an additional discount of 2% would be allowed. Assuming the bill was paid in 30 days, how much did the ink cost including the shipping charge of $8.45?   _____

13. A long-time customer of Brunelda Binding and Shipping Services receives a standing discount of 10% on each job that is ordered, assuming that a minimum of $2,500 worth of binding and finishing work is purchased each year. In addition, the customer receives a 2½% discount for paying his bills within 30 days. What is the bottom-line cost of a job priced out at $1,435.73 if the stated conditions are met?   _____

14. The paper stock and labor costs for an order of letterheads and envelopes amounted to $535.64. The graphic arts company business manager added the usual 18% for company profit and overhead. If 54% of the cost and profit were allocated only to the letterheads, what was the billing price for this portion of the job?   _____

15. An author of children's books received an invoice for $3,500 due in 30 days, but there was a 2% discount if the invoice was paid immediately in cash (Figure 20–2). The author did not wish to remove this amount from his savings account, thus he decided to borrow the $3,500 for 30 days at 10% annual interest so the invoice could be paid in cash. How much money did the author save by borrowing the money and immediately paying the invoice?   _____

Figure 20–2 Many children's books are published each year that are written by authors around the world.

Measurement

SECTION

5

Measurement is the size of something. For example, the playing area of a football field is 100 yards long and 50 yards wide. Over the years, several measurement systems have been developed by different people for different reasons. The *customary measurement system* is the standard system of inches, feet, yards, quarts, gallons, and so on. It is widely used throughout the United States but has limited use throughout most of the other countries of the world. The *point measurement system* was developed for use by typesetters and other people involved in the graphic arts industry who needed a precise method of measurement for small to large type characters. There are approximately 72 *points* per customary inch. The word *approximately* is used because 72 points do not equal an exact inch, but 72 points are closer to one inch than 71 or 73 points.

The *metric measurement system* is based on the decimal system of 10s. There are ten parts to each of the submeasurement groups in the metric system. There is no direct relationship of the metric system to the customary system or the point system. It is a complete measurement system that is used in the scientific arena and in many industrial areas. People throughout most countries of the world utilize this system as their base system of measurement except for the United States.

Ratio and *proportion* are significant measurement entities in themselves. Ratio means that something (a measurement) is a given part of another something. A ratio can be expressed as one number to another; for example, 8 to 4 means that 8 is twice that of 4, while 4 to 8 means that 4 is ½ of 8. Proportion is a part of ratio; for example, 5% of 100 is 20. Making something 140% of a known measurement means that the result will be 100% plus 40% of the original. For example, 140% of 90 is equal to 90 + 36, or 126.

The *micrometer* is a precision measuring device, a tool, just as a ruler, tape measure, or framing square is a measuring device. The micrometer permits measurements in thousandths of an inch.

Unit 21 CUSTOMARY MEASUREMENT SYSTEM

BASIC PRINCIPLES OF CUSTOMARY MEASUREMENT SYSTEM

There are four categories or groups of the customary measurement system that are important for people in graphic communications to know. These four measurement groups are often used in both the technical and management/supervisory areas of graphic communications. The four measurement groups are *linear*, *liquid*, *weight*, and *area*. The basic principles of each are listed in this unit for reference and study before the problems provided in this unit are encountered.

Linear Measurement

This measurement group involves the determination of length by using such standard measures as inch, foot, and yard. Examples include the following:

Examples: 1 inch = two halves or ½s
  = four quarters or ¼s
  = eight eighths or ⅛s
  = sixteen sixteenths or 1/16s

| 12 inches (in) = 1 foot (ft) | inches ÷ 12 = feet |
| | feet × 12 = inches |

| 3 feet (ft) = 1 yard (yd) | feet ÷ 3 = yards |
| | yards × 3 = feet |

| 36 inches (in) = 1 yard (yd) | inches ÷ 36 = yards |
| | yards × 36 = inches |

Liquid Measurement

This measurement group involves the determination of the amount of liquid within a given space using such standard measures as fluid ounce, cup, pint, quart, and gallon. Examples include the following:

Examples: 8 fluid ounces (fl oz) = 1 cup   ounces ÷ 8 = cups
  cups × 8 = ounces

2 cups = 1 pint (pt)   cups ÷ 2 = pints
  pints × 2 = cups

2 pints (pt) = 1 quart (qt)   pints ÷ 2 = quarts
                                   quarts × 2 = pints

4 quarts (qt) = 1 gallon (gal)   quarts ÷ 4 = gallons
                                   gallons × 4 = quarts

Weight Measurement

This measurement group involves the determination of how heavy something is and includes such standard measures as ounce and pound. Examples include the following:

Examples: 16 ounces (oz) = 1 pound   ounces ÷ 16 = pounds
                                             pounds × 16 = ounces

Area Measurement

This measurement group involves the determination of the surface of a given space in square units. The square units commonly used are the square inch and the square foot. Examples include the following:

Examples: The square inch is a square space that measures 1 inch on each side; and the square foot is a square space that measures 1 foot on each side (Figure 21–1).

Figure 21–1 Equal measurements on each side are peculiar to both the square inch and the square foot.

Since there are 12 inches in 1 foot, each side of a square foot is 12 inches long; therefore, there are 144 square inches in a square foot (Figure 21–2).

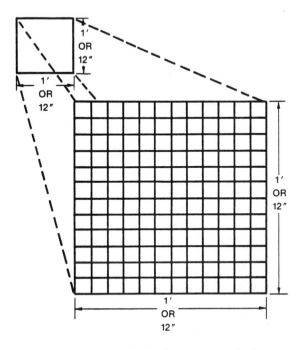

Figure 21–2   There are 144 square inches in one square foot.

To find the area of a rectangle, it is necessary to multiply the width times the length. Two examples are provided for study and reference.

Example 1: To find the area of a rectangle 3½" wide and 5½" long, simply multiply one measure by the other.

Area = width × length

Area = 3½" × 5½" = 3.5 × 5.5 = 19.25 sq in

Example 2: To find the area of two rectangles of different sizes that are tied together, establish the dimensional measurements of each, find the squared area of each in either square inches or square feet, and add the two squared amounts together (Figure 21–3).

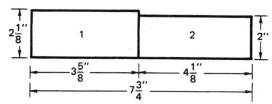

Figure 21–3 The squared area of joined rectangles is determined by calculating the squared area of each rectangle and then adding the two squared amounts together.

Area of rectangle 1 = 2⅛" × 3⅝" = 2.125 × 3.625 = 7.703125 sq in

Area of rectangle 2 = 2" × 4⅛" = 2" × 4.125" = 8.25 sq in

Area of rectangles 1 and 2 = 7.703125 sq in + 8.25 sq in = 15.953125 sq in

PRACTICAL PROBLEMS

Note: In problems 1–10, determine the indicated measurements on the foot rule (ruler) as identified in the illustration (Figure 21–4):

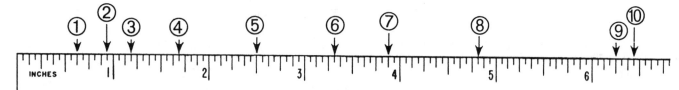

Figure 21–4 The typical foot rule contains measurements of inch, half-inch, quarter-inch, eighth-inch, and sixteenth-inch.

1. _____

2. _____

3. _____

4. _____

5. _____

6. _____

7. _____

8. _____

9. _____

10. _____

Note: In problems 11–20, use the illustration (Figure 21–5) and a foot rule to measure the lengths as indicated to the nearest eighth of an inch. All fractions should be expressed in the lowest terms.

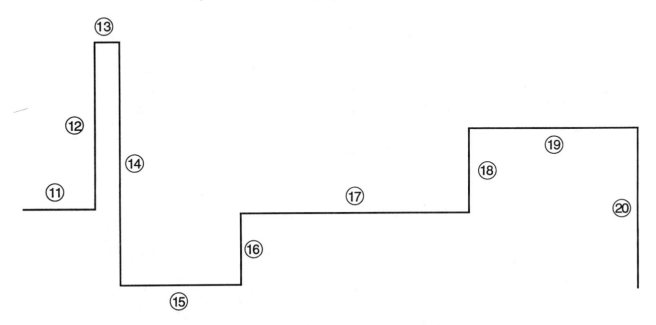

Figure 21–5   A measurement exercise that is to be measured to the nearest ⅛ inch.

11. _____

12. _____

13. _____

14. _____

15. _____

16. _____

17. _____

18. _____

19. _____

20. _____

Note: In problems 21–30, use the illustration (Figure 21–6) and a foot rule to measure the lengths as indicated to the nearest sixteenth of an inch. All fractions should be expressed in the lowest terms.

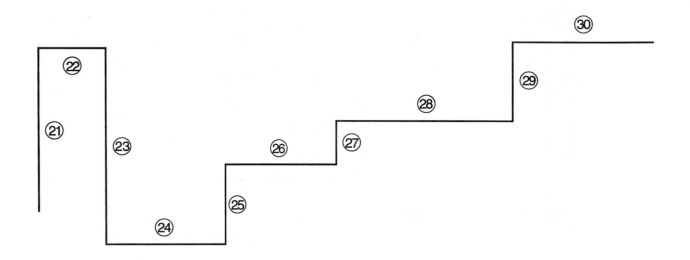

Figure 21–6   A measurement exercise that is to be measured to the nearest ¹⁄₁₆ inch.

21. _____	26. _____
22. _____	27. _____
23. _____	28. _____
24. _____	29. _____
25. _____	30. _____

Note: In problems 31–40, express each initial measure in the indicated equivalent measurement.

31. 3'7" as inches   _____

32. 7" as decimal feet   _____

33. 6'8" as inches   _____

34. 42" as decimal yards   _____

35. 2.34 yards as feet   _____

36. 6.45' as decimal inches   _____

37. $9\frac{3}{4}$" as decimal feet   _____

38. 3.45' as decimal inches   _____

39. 4'8" as decimal yards   _____

40. $17\frac{1}{8}$" as decimal feet   _____

Note: In problems 41–60, express the measures provided in the indicated equivalent measurements.

41. _____ pints = 1 gallon

42. _____ cups = 1 gallon

43. _____ ounces = 1 gallon

44. _____ gallons = 8 pints

45. _____ cups = 9 quarts

46. _____ ounces = 4 gallons

47. _____ cups = $3\frac{3}{4}$ gallons

48. _____ ounces = $7\frac{1}{2}$ quarts

49. _____ quarts = $6\frac{3}{4}$ cups

50. _____ gallons = 2.63 cups

51. _____ ounces = 7.83 gallons

52. _____ quarts = 47 cups

53. _____ cups = $6\frac{2}{3}$ quarts

54. _____ gallons = 7.625 pints

55. _____ cups = 4.63 quarts

56. _____ pints = 47.63 ounces

57. _____ cups = .34 gallon

58. _____ ounces = $8\frac{1}{2}$ quarts

59. _____ quarts = 7.63 gallons

60. _____ gallons = 47.8 pints

Note: In problems 61–65, express the measures provided in the indicated equivalent measurements.

61. $2\frac{3}{4}$ ounces = _____ pound

62. _____ ounces = $17\frac{3}{4}$ pounds

63. 17.78 pounds = _____ ounces

64. _____ pounds = $18\frac{2}{3}$ ounces

65. _____ ounces = $25\frac{3}{4}$ pounds

66.   Determine the square inches in a rectangle 2½" × 3½".   _____

67.   Determine the square inches in a rectangle 6¼" × 8¾".   _____

68.   Determine the square inches in a rectangle 20½" × 24¾".   _____

69.   Determine the square feet in a rectangle 3'4" × 7'8".   _____

70.   Determine the square feet in a rectangle 4'8¾" × 5'7½".   _____

71.   Determine the square inches in the rectangle in Figure 21–7.   _____

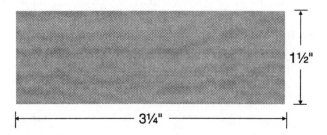

Figure 21–7   How many square inches are there in this rectangle?

72.   Determine the square inches in the rectangle in Figure 21–8.   _____

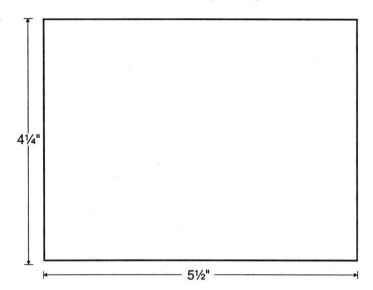

Figure 21–8   How many square inches are there in this rectangle?

73. Determine the square inches in the L-shaped rectangle in Figure 21–9.   _____

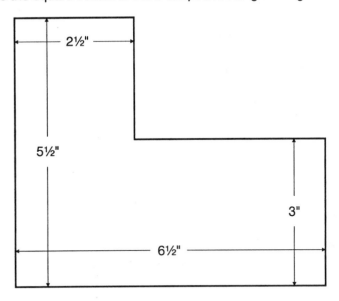

Figure 21–9   The square inches in this L-shaped rectangle can be determined two ways.

74. Determine the square inches in the U-shaped rectangle in Figure 21–10.   _____

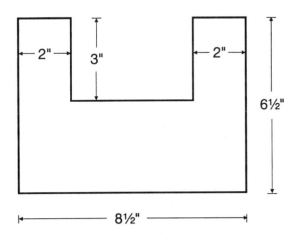

Figure 21–10   As with Figure 21–9, the square inches in this U-shaped rectangle can also be determined two ways.

75.   Determine the square inches in the gray shaded area minus the white area
      in the center (Figure 21–11).   _____

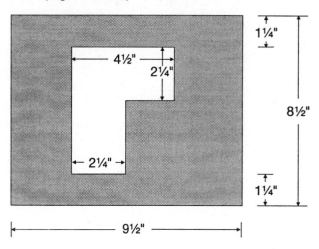

Figure 21–11   Previously learned skills in calculating square inches will be needed in
determining the square inches in the gray area only.

Unit 22   POINT MEASUREMENT SYSTEM

BASIC PRINCIPLES OF POINT MEASUREMENT SYSTEM

The American point system was adapted in the United States in 1886 by members of the United States Typefounders Association. This system has two units of measure known as the *point* and the *pica*. In reference to the customary inch, the point measures 0.013832 inch or approximately ¹⁄₇₂ of an inch. Points are used to measure the height of typefaces and vertical line spacing. The second unit of measure in the point measurement system is the pica. This unit of measure equals approximately ¹⁄₆ of an inch and includes exactly 12 points. The pica measurement is used to determine lengths of type lines and depths of type and other content on a page. In summary, the comparison of points and picas to the customary inch are listed as follows:

72 points   =   1 inch (approximately)

12 points   =   1 pica (exactly)

6 picas   =   1 inch (approximately)

6 points   =   one-half (½) pica (exactly)

1 point   =   ¹⁄₇₂ inch (approximately) (Figure 22–1)

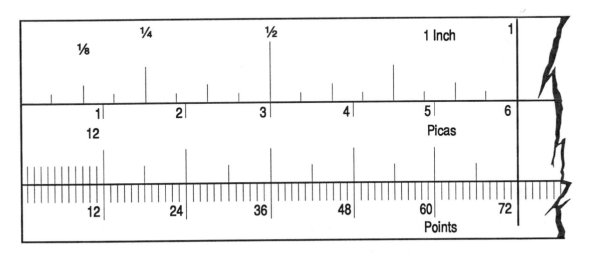

Figure 22–1 A portion of a measuring rule showing the comparison among inches, picas, and points.

A *line gauge* is the common measuring device used to measure type lines, images, and pages in points, half-picas, picas, and inches (Figure 22–2). Line gauges are generally made of steel or aluminum and are commonly available in 12- to 18-inch lengths. It is somewhat difficult to precisely measure points with a line gauge, but most of the time it is possible to come very close to the exact measurement. Remembering that half-picas contain only six points, it is rather easy to assess the length, height, or thickness of some object or image in picas and points using a line gauge.

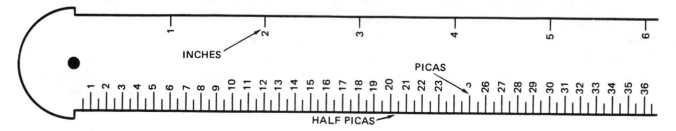

Figure 22–2   A portion of a line gauge showing the relationship of inches to picas and half-picas.

Converting Points, Picas, and Inches

Even though points and picas are not perfectly compatible with inches, it is often necessary to convert these measures back and forth. The measures and conversions are usually close enough to not cause problems with the positioning of type, illustrations, and pictures. Obviously, the conversions must be mathematically correct, thus the need for knowing the following formulas:

> known inches × 72 = points
> known picas × 12   = points
> known points ÷ 12   = picas
> known inches × 6   = picas
> known points ÷ 72   = inches
> known picas ÷ 6   = inches

The problems in the following examples illustrate these conversions.

Example 1:   The depth of some copy is 2¾ inches. What is the depth of the copy in points?

  Solution = 2¾ inches or 2.75 × 72 points = 198 points

Example 2:   A type page measures 39 picas wide. What is the width of the form in inches as well as in points?

  Solution = 39 picas ÷ 6 picas (per inch) = 6.5 or 6½ inches and 39 picas × 12 points (per pica) = 468 points

Adding, Subtracting, Multiplying, and Dividing Points and Picas

When working with picas and points, it is valuable to express the picas and points entirely as points and then convert back to picas and points as needed after the calculations have been made. For example, if something measures 14 picas and 6 points, it is convenient to work with 174 points when performing any of the basic mathematical functions. Adding 14 picas and 6 points to 19 picas and 9 points is easy when these measurements are converted to 174 points and 237 points. These two numbers total 411 points; and when the total is reconverted to picas and points, the amount is 34 picas and 3 points.

PRACTICAL PROBLEMS

1. The measurements of a newspaper advertisement are 4" × 3¾" (Figure 22–3). What are the measurements of the advertisement in picas? _____

Figure 22–3 An advertisement for a brief vacation designed for the entertainment sections of daily newspapers.

2. The depth of some copy is 4¾ inches. What is the depth of the copy in picas? _____

3. An inch is equal to approximately 72 points. What part of an inch is 6 points? _____

4. A page of type measures 288 points wide. What is the width in inches? _____

5. An inch is equal to approximately 72 points. What fraction of an inch is 54 points? _____

6. A line of body type measures 17 picas long. What is the decimal length in inches?

    _____

7. A line of headline type is 42 picas long. What is its length in points?

    _____

8. A body of type measures 5⅞" wide × 5⅝ deep. What are the width and depth in points?

    _____

9. A job is printed on stock cut 8½" wide × 11" deep. What are the width and depth in picas?

    _____

10. For a business form, the computer copy preparation specialist has to create 114 lines with each being 216 points long. How many total inches of lines are created for this job?

    _____

11. A series of copy measurements includes: 3 picas and 3 points; 12 picas; 12 picas and 3 points; and 12 picas and 6 points. What is the total amount of these measurements in picas and points?

    _____

12. Another series of copy measurements includes: 11 picas and 4 points; 16 picas and 10 points; and 21 picas and 8 points. What is the total amount of these measurements in picas and points?

    _____

13. A third series of copy measurements includes: 23 picas and 9 points; 45 picas and 3 points; and 19 picas and 7 points. What is the total amount of these measurements in picas and points?

    _____

14. One line of type is 15 picas and 9 points long while a second line is 34 picas and 7 points long. How much longer is the second line than the first line in picas and points?

    _____

15. On another page, the computer copy preparation artist keyboards two lines of type that are 23 picas and 8 points long and 63 picas and 3 points long. How many points longer is the second line than the first line?

    _____

16. There is a caption under a figure in a publication that includes two lines of type. One line is 45 picas and 2 points long while the second line is 32 picas and 5 points long. How much shorter is the second line than the first line in picas and points?

    _____

17. A type form includes eight columns of type characters (letters and numbers). Each of the columns is established at 3 picas and 4 points wide (Figure 22–4). What is the total width of the eight-column type form in picas and points? _____

ALPHABET LENGTH IN POINTS	CHAR. PER PICA	ALPHABET LENGTH IN POINTS	CHAR. PER PICA	ALPHABET LENGTH IN POINTS	CHAR. PER PICA	ALPHABET LENGTH IN POINTS	CHAR. PER PICA
60 PTS	5.70	102 PTS	3.35	144 PTS	2.38	186 PTS	1.84
61 PTS	5.61	103 PTS	3.32	145 PTS	2.36	187 PTS	1.83
62 PTS	5.52	104 PTS	3.29	146 PTS	2.34	188 PTS	
63 PTS	43	105					

Figure 22–4 An eight-column type form that includes type characters of both letters and numbers.

18. Another type form includes nine columns of type characters. Each of the columns is 2 picas and 10 points wide. What is the total width of the nine columns in points? _____

19. Type characters are often measured in an average *set-width*. In counting the number of type characters in a long line of type, the layout artist determines that there are 81 type characters in a line. If the average set-width of each type character is 6 points, how long is the line of type in picas and points, not considering the word spaces? _____

20. A page of type is to be typeset in a 45-pica width space. How many columns of type 3 picas and 9 points wide can be fit into the 45-pica width space? _____

21. A headline in a company newsletter is to be 27½ picas long, and the capital letters that are to be used have an average set-width of 2½ picas. How many type characters long can the headline be? _____

22. On a page of a financial report, there are to be eight columns of figures that have to be typeset within the 42-pica width space. How wide in points will each column need to be so the first and eighth columns touch both sides of the space? _____

23. Three lines of type are each 15 picas and 8 points long. Two other lines are each 52 picas and 5 points long. How long in picas and points are the five lines of type if they are set end to end?

   _____

24. Six lines of type are each 34 picas and 9 points long. It is necessary to reduce each line by 9 picas and 3 points. What is the total length of the revised six lines of type in picas and points?

   _____

25. There are a total of 20 lines of 11-point type in two paragraphs. The typesetter decides to include 3 points of *leading* between each line. What is the total depth of the type form that includes the leading between the 20 lines?

   _____

 **Unit 23** *METRIC MEASUREMENT SYSTEM*

BASIC PRINCIPLES OF METRIC MEASUREMENT SYSTEM

The metric measurement system was first proposed in the year 1670 by Gabriel Mouton, a Frenchman. Through the years, many improvements have been made in the metric system. For example, in 1875 an international treaty called the "Treaty of the Metre" was established to set up metric standards for length and weight. The treaty was signed by representatives of seventeen countries, including the United States of America. By 1900, a total of thirty-five worldwide nations had officially accepted the metric system. Today, except for the United States and a few smaller countries, the entire world is utilizing the metric system as the base measurement system. Irregardless of the limited use of metrics in the United States, considerable use of this measurement system exists in the photographic, scientific, and medical areas. There are four measurement groups of the metric system that graphic communications people need to know. These four groups permit people to make measurements referred to as *linear*, *liquid*, *weight*, and *area*.

Linear Measurement

The *meter* is the standard unit of linear measurement in the metric system, and it is slightly longer than the customary yard. The meter (m) is subdivided into 10 equal parts called *decimeters* (dm), 100 equal parts called *centimeters* (cm), and 1,000 equal parts called *millimeters* (mm). It is very easy to change from one metric linear measure to another, as shown in the following examples.

Examples: 10 millimeters (mm) = 1 centimeter (cm)   mm ÷ 10 = cm
                                                            cm × 10 = mm

                10 centimeters (cm) = 1 decimeter (dm)   cm ÷ 10 = dm
                                                            dm × 10 = cm

                10 decimeters (dm) = 1 meter (m)   dm ÷ 1 = m
                                                            m × 10 = dm

In actuality, decimals are moved to the left or right without having to multiply or divide by ten, thus conversions can be rapidly made from millimeter to centimeter to decimeter and vice versa.

Examples: 1.000 = one meter

                10.00 = one decimeter

                100.0 = one centimeter

                1000.0 = one millimeter

Liquid Measurement

The *liter* is the standard unit of liquid measurement in the metric system, and it is slightly larger than the customary quart. The liter (L) is subdivided into 1,000 equal parts called *milliliters* (mL). It is very easy to change from liter to milliliter or milliliter to liter, as shown in the following example.

Example: 1 liter (L) = 1,000 milliliters (mL)   mL ÷ 1,000 = L

L × 1,000 = m

The mathematical functions involve either multiplication or division, but as with linear measure, it is easy to move the decimal three places to the left or to the right; for example, 1.000 liter or 1,000.0 milliliters.

Weight Measurement

The *gram* is the standard unit of weight measurement in the metric system, and it is approximately the weight of a medium-sized paper clip. Since the gram (g) is such a small unit of weight, the *kilogram* (kg) is used for larger objects. With kilo meaning 1,000 times, there are 1,000 grams in one kilogram. In changing from kilogram to gram or from gram to kilogram, division or multiplication by 1,000 must be done, as shown in the following example.

Example: 1 kilogram (kg) = 1,000 grams (g)   g ÷ 1,000 = kg

kg × 1,000 = g

Area Measurement

The area of a surface is measured in square units, and the commonly used unit is the *square centimeter*, which measures one centimeter on each side (Figure 23–1).

Figure 23–1   There are equal measurements on each side of the square centimeter.

The square centimeter is often abbreviated as cm^2. Since there are ten millimeters in each centimeter, there are 100 square millimeters (mm^2) in each square centimeter (cm^2) (Figure 23–2).

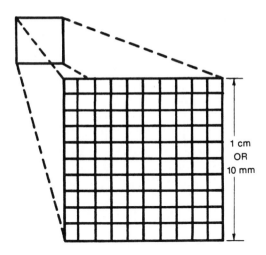

Figure 23–2   There are 100 square millimeters in one square centimeter.

Conversion of Measurements

Most of the time, people in graphic communications are working in either the metric system or the customary system for typical measurements. This is true except when they are working with type matter, and then the point measurement system is used. There are times when it becomes necessary to change from the metric to the customary system and vice versa. When conversion is necessary, the appropriate formula should be used for both convenience and accuracy, as shown below:

1 inch = 2.54 centimeters or 25.4 millimeters

1 centimeter = .3937 inch	cm = 2.54 × in
1 quart = .946 liter or 946 milliliters	in = .3937 × cm
1 liter = 1.057 quarts	L = qt × .946
1 pound = .54 kilogram or 454 grams	qt = L × 1.057
1 kilogram = 2.204 pounds	kg = .454 × lb
mm = 25.4 × in	lb = kg × 2.204

Trying to compare metric measurements to customary measurements and vice versa is very difficult and not recommended. If conversions must be done, it is best to utilize the appropriate conversion factors and perform the needed mathematical calculations.

PRACTICAL PROBLEMS

Note: In problems 1–10, determine the indicated measurements on the metric rule in centimeters and millimeters as identified in the illustration (Figure 23–3).

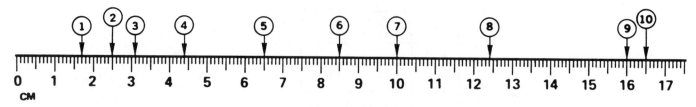

Figure 23–3   The typical meter rule contains measurements of centimeters and millimeters.

1. _____   6. _____

2. _____   7. _____

3. _____   8. _____

4. _____   9. _____

5. _____   10. _____

Note: In problems 11–20, place the measurements given in the problems on the metric rule in Figure 23–4 using the same marking method as that used in Figure 23–3.

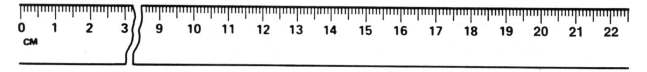

Figure 23–4   What measurements are where on this meter rule?

11.   13 cm   16.   0.9 cm

12.   16.5 cm   17.   17 mm

13.   13 mm   18.   19.3 cm

14.   165 mm   19.   147 mm

15.   9.6 cm   20.   108 mm

Note: In problems 21–30, use the illustration (Figure 23–5) and a metric rule to measure the lengths in centimeters as indicated.

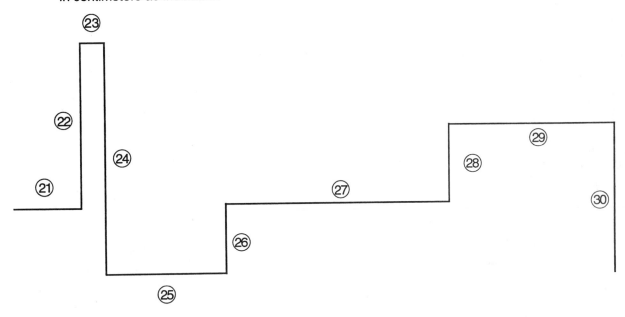

Figure 23–5 A measurement exercise that is to be measured in centimeters.

21. _____   26. _____

22. _____   27. _____

23. _____   28. _____

24. _____   29. _____

25. _____   30. _____

Note: In problems 31–40, use the illustration (Figure 23–6) and a metric rule to measure the lengths in millimeters as indicated.

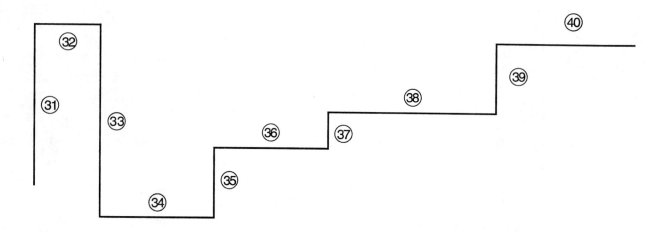

Figure 23–6   A measurement exercise that is to be measured in millimeters.

31. _____   36. _____

32. _____   37. _____

33. _____   38. _____

34. _____   39. _____

35. _____   40. _____

Note: In problems 41–50, express each given measure in the indicated equivalent measurement.

41.   7.3 cm = _____ mm

46.   47 mm = _____ cm

42.   75 mm = _____ cm

47.   _____ mm = 57.3 cm

43.   17.8 cm = _____ mm

48.   3.2 cm = _____ mm

44.   _____ mm = 19.3 cm

49.   98.5 cm = _____ mm

45.   _____ cm = 153 mm

50.   _____ mm = 153 cm

Note: In problems 51–60, express each given measure in the indicated equivalent measurement.

51.   65 mL = _____ L

56.   457 mL = _____ L

52.   0.065 L = _____ mL

57.   3.5 L = _____ mL

53.   2 L = _____ mL

58.   _____ mL = 1.8 L

54.   _____ L = 465 mL

59.   _____ mL = 1.8 L

55.   _____ mL = 2,345 L

60.   0.2 L = _____ mL

Note: In problems 61–70, express each given measure in the indicated equivalent measurement.

61.   3 kg = _____ g

66.   458 g = _____ kg

62.   3.7 kg = _____ g

67.   _____ g = 9.3 kg

63.   _____ kg = 4,300 g

68.   _____ kg = 750 g

64.   _____ g = 0.83 kg

69.   _____ g = 0.085 kg

65.   96 g = _____ kg

70.   0.375 kg = _____ g

71.   Determine the area in a rectangle that is 18 cm × 15 cm.   _____

72.   Determine the area in a rectangle that is 19.3 cm × 18.6 cm.   _____

73.   Determine the area in a rectangle that is 23.7 cm × 35.8 cm.   _____

74.   Determine the area in a rectangle that is 238 mm × 465 mm and provide the
      answer in cm².

75.   Determine the area in a rectangle that is 455 mm × 685 mm and provide the
      answer in cm².

76.   Determine the square centimeters in the rectangle in Figure 23–7.

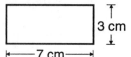

Figure 23–7   How many square centimeters are there in this rectangle?

77.   Determine the square millimeters in the rectangle in Figure 23-8.

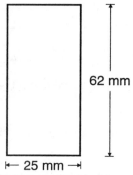

Figure 23–8   How many square millimeters are there in this rectangle?

78.   Determine the square centimeters in the L-shaped illustration in Figure 23–9.

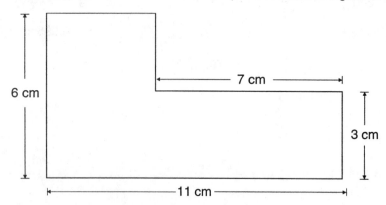

Figure 23–9   The square centimeters in this L-shaped illustration can be determined two ways.

79. Determine the square millimeters in the U-shaped illustration in Figure 23–10. _____

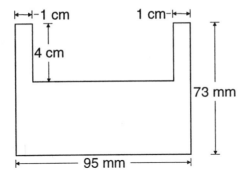

Figure 23–10   As with Figure 21–9, the square millimeters in this U-shaped illustration can be determined two ways.

80. Determine the square centimeters in the gray shaded area minus the white area in the center (Figure 23–11). _____

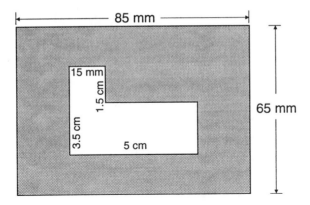

Figure 23–11   Previously learned skills in calculating square centimeters will be needed in determining the square centimeters in the gray area only.

Note: In problems 81–90, express each given measure in the indicated equivalent measurement.

81. 14½ in = _____ cm

82. 6.5 cm = _____ in

83. _____ cm = 2'9"

84. 4'7" = _____ m

85. _____ qt = 453 mL

86. 4.5 L = _____ qt

87. _____ cups = 1.5 L

88. 4 fluid ounces = _____ mL

89. 5½ lb = _____ kg

90. _____ lb = 475 g

Unit 24   CALCULATING RATIO AND PROPORTION

BASIC PRINCIPLES OF CALCULATING RATIO AND PROPORTION

Ratio and proportion are important mathematical principles for people in graphic communications to know. There are times when it is necessary to determine the relationship of one numerical amount to another so that the numbers can be fully understood. Knowing how to determine page proportions and what the width measurement is to the height measurement may help someone design a better printed page.

Ratio

A ratio is the relationship between two quantities of the same kind. A ratio can be expressed in four ways: (1) in words, as in the ratio of 12 to 4; (2) by use of a colon, as in 12:4; (3) by use of a division sign, as in 12 ÷ 4; or (4) as a fraction, as in 12/4. The fractional method is, perhaps, the easiest form to work with and use. A ratio should always be expressed in the lowest terms, as shown in the following examples:

Examples: $8/12 = 2/3$

10 to 15 = 2 to 3

8:4 = 2:1

The ratio of the father's weight to the son's weight is 180 to 75.

$180/75 = 36/15 = 12/5$

The ratio of the son's weight to the father's weight is 75 to 180.

75 to 180 = 5 to 12. (It does matter which number comes first. Keep in mind what is being compared to what.)

In order to find the ratio of the father's height to the son's height, it is necessary to express both heights in the same unit. The ratio of the father's height of 72 inches to the son's height of 52 inches = 18 to 13 or 6 feet to 4⅓ feet.

$6 ÷ 4⅓$ feet $= 6/1 ÷ 13/3 = 6/1 × 3/13 = 18/13$

Percent as a Ratio

Percent is an application of a ratio. This is shown in the following examples.

Examples: 6% means 6 to 100 = $\frac{6}{100}$ = $\frac{3}{50}$

5% means 5 to 100 = $\frac{5}{100}$ = $\frac{1}{20}$

53% means 53 to 100 = $\frac{53}{100}$

$3\frac{1}{3}$% means $3\frac{1}{3}$ to 100 = $3\frac{1}{3}$ ÷ 100 = $\frac{10}{3}$ ÷ $\frac{100}{1}$ = $\frac{10}{3}$ × $\frac{1}{100}$ = $\frac{10}{300}$ = $\frac{1}{30}$

Proportion

A *proportion* is the equality of two (or more) ratios. A proportion can be expressed in three ways: (1) 3:4::9:12; (2) 3:4 = 9:12; and (3) $\frac{3}{4}$ = $\frac{9}{12}$. In each case, the expression is read as follows: 3 is to 4 as 9 is to 12. The fractional form (number 3) is utilized in this textbook to solve problems. To solve a proportion, it is necessary to cross multiply and then solve for the unknown factor, as shown in the following examples.

Examples: $\frac{3}{7} = \frac{N}{14}$

$N = (3 \times 14) \div 7 = 42 \div 7 = 6$

$\frac{N}{6} = \frac{5}{3}$

$N = (5 \times 6) \div 3 = 30 \div 3 = 10$

$\frac{5}{8} = \frac{6}{N}$

$N = (8 \times 6) \div 5 = 48 \div 5 = 9\frac{3}{5}$

$\frac{2\frac{1}{2}}{N} = \frac{7}{6}$

$N = (2\frac{1}{2} \times 6) \div 7 = 15 \div 7 = 2\frac{1}{7}$

PRACTICAL PROBLEMS

Note: In problems 1–10, express the ratios in lowest terms.

1. 4 to 16   _____   6.   5:2½   _____

2. 6 to 9   _____   7.   .25:.635   _____

3. 15 to 5   _____   8.   3 inches:1 foot   _____

4. 32 to 20   _____   9.   52 minutes to 2 hours   _____

5. 2½:10   _____   10.   45¢ to $5   _____

Note: In problems 11–14, express the percents as ratios in lowest terms.

11. 75% _____ 13. 2½% _____

12. 4% _____ 14. 115% _____

Note: In problems 15–22, solve for *N*.

15. $N/5 = 7/10$ _____ 19. $3.4/8 = N/7$ _____

16. $8/9 = N/15$ _____ 20. $N/6.3 = 5/8$ _____

17. $7/N = 5/6½$ _____ 21. $6/N = 10/5$ _____

18. $3¾/7 = 5/N$ _____ 22. $N/2½ = 3/5.7$ _____

23. An experienced flexography press operator earns $525 per week and the assistant press operator earns $300 per week. What is the ratio of the assistants's wage to the press operator's wage? _____

24. The dimensions of a picture are 5" × 7". A photographer wants to enlarge it to an 8" × 10" size. What is the ratio of the width of the picture to the width of the enlargement? **Note:** The enlargement will not maintain the proportion of the original picture. _____

25. A photograph that will be used in a newspaper column measures 6" × 8". What height will the photograph be in picas if the newspaper column is 12 picas wide? _____

26. An original photograph that was 8" wide and 4" high must be made to fit into a space that is 24 picas wide. What must the depth of the photograph be to accommodate the 24-pica width and keep the same proportion? _____

27.   A photograph that is 5" wide and 8" high must fit into an 18-pica width on the editorial page of the city newspaper (Figure 24–1). How deep in picas must the photograph be to keep the same proportion as the original photograph?   _____

Figure 24–1   Often original photographs must be kept in the same proportion in publications even when they are enlarged or reduced in size.

28.   Some type copy that was originally set for a 24 pica wide by 18 pica high space must be reset to fit a type page that is 36 picas wide. What is the required new height of the type page space?   _____

29.   A good customer requests a specific shade of yellow ink to be used on a job that she has ordered. From past experience, the litho press operator determines that 3¾ ounces yellow, ¼ ounce green, and 12 ounces of mixing white are needed to make one pound of yellow ink the exact color the customer desires. The press operator also estimates that to run this particular job, it will require 3½ pounds of ink. How many ounces of the mixing white ink will be needed to make the 3½ pound batch of ink?   _____

30.   In order to match a medium blue ink that had been used on the previous run of a high school music event flyer, a litho press operator used the following formula: six parts of reflex blue, two parts of cyan, and eight parts of mixing white. How many ounces of reflex blue were used to make the needed ½ pound of the medium blue ink to run the job?   _____

Unit 25   USING THE MICROMETER

BASIC PRINCIPLES OF USING THE MICROMETER

The micrometer, sometimes nicknamed "mike," is an instrument used to accurately measure objects to a thousandth of an inch. Micrometers are available in various sizes, but graphic communications personnel usually do not need one larger than the one-inch size (Figure 25–1).

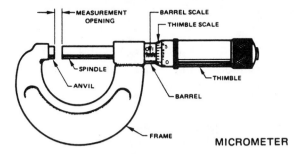

Figure 25–1   The parts of a standard micrometer. To obtain more accurate readings, micrometers with larger anvils and spindle ends are sometimes used for graphic communications purposes.

The one-inch size means that the maximum measurement opening between the *anvil* and the *spindle* is one inch. Most of the time, people in graphic communications use micrometers to measure the thickness of paper. Litho press operators use these precise instruments to measure the thickness of plates, blankets, and the paper packing used on the cylinders of their presses. Platemaking personnel and press operators of other printing processes such as flexography, gravure, and screen also find excellent use for micrometers.

The digits on the barrel of the micrometer represent .100 inch, .200 inch, .300 inch, and so on (Figure 25–2). The four subdivisions between the digits each represent .025 inch. Each turn of the thimble moves it from one division line to the next, or .025 inch. The numbers along the thimble indicate thousandths of an inch.

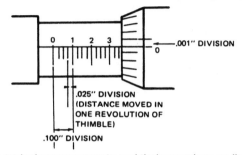

Figure 25–2   The typical measurements and their meanings as listed on the barrel scale of the micrometer.

To read a micrometer, or "mike," the number of hundreds of thousandths must first be determined. Once done, the number of .025-inches are determined and added (Figure 25–3).

.200
.050
―――
.250"

Figure 25–3 An example of the thimble position when measuring a material that is 250 thousandths of an inch thick.

After that, the thousandths of inches as identified on the thimble are added to create the total thickness. An example of the thimble position is illustrated in Figure 25–4. Remember, the micrometer is a precision measurement tool and should be handled with care.

.500
.050
.012
―――
.562"

Figure 25–4 An example of the thimble position when measuring a material or group of materials that is 562 thousandths of an inch thick.

PRACTICAL PROBLEMS

Note: In problems 1–10, study the ten micrometer thickness settings and provide the indicated reading on each micrometer image (Figure 25–5). Remember to show the decimal and to use the inch (") designation on each of the ten responses.

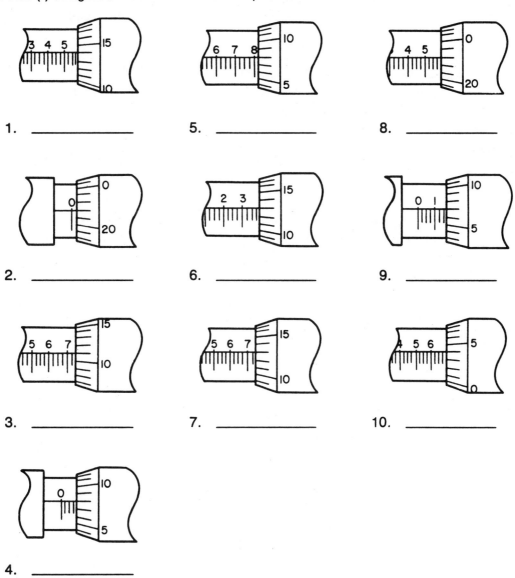

1. _____

5. _____

8. _____

2. _____

6. _____

9. _____

3. _____

7. _____

10. _____

4. _____

Figure 25–5   Identify these ten barrel scale and thimble scale micrometer thickness readings.

11. Using a micrometer, measure the thickness of a sheet of 20-pound, white bond paper provided by your instructor. What is the thickness of this bond paper in thousandths of an inch? _____

12. Using a micrometer, measure the thickness of a sheet of 50-pound, white book paper provided by your instructor. What is the thickness of this book paper in thousandths of an inch? _____

13. Using a micrometer, measure the thickness of a sheet of 65-pound, red or blue cover paper, antique finish, provided by your instructor. What is the thickness of this cover paper in thousandths of an inch? _____

14. Using a micrometer, measure the thickness of a sheet of 110-pound, white index/bristol paper provided by your instructor. What is the thickness of this index/bristol paper in thousandths of an inch? _____

15. Using a micrometer, measure the thickness of a lithographic press blanket provided by your instructor. What is the thickness of the litho press blanket in thousandths of an inch? _____

Illustrations and Photographs

Visual images in the form of *illustrations* and *photographs* are both informative and entertaining. When reading a newspaper or book or when looking at a personal or company web page, illustrations and photographs are utilized to help inform you about the content. Sometimes illustrations are in the form of artwork, drawn either by hand or with the aid of a computer. These illustrations can be very basic or very complex, depending on the need and how much communication needs to be conveyed. The same is true with photographs; they are used in graphic communication products to help in conveying the overall message that is being communicated via words, sentences, and paragraphs.

Graphs and *charts* are considered to be illustrations in that they are created to provide information that will be important to the reader. There are several kinds of graphs and charts presented in Unit 24. It can be safely stated that graphs and charts are used to inform readers of specific information rather than entertain them.

Photographs, on the other hand, are used in graphic products to both inform and entertain. Who doesn't carry one or more photographs with them in purses and billfolds? Printing photographs in publications provides considerable information and, much of the time, a vast amount of entertainment for readers and nonreaders alike. Take, for example, a child who does not know how to read but can look at the printed pictures and gain considerable enjoyment.

Sizing of both illustrations and photographs is important to the overall imaged substrate. Images that are too small tend not to communicate, thus they must be correctly sized on the graphically produced page. In addition, illustrations and photographs can be reproduced too large so they do not correctly fit on a page or are difficult to use because the large size is more than the human eye can really see at one time. Knowing how to size illustrations and photographs by both hand and computer methods greatly enhances the value of designers and copy preparation personnel. Both sizing methods are presented in Units 25 and 26; in addition, several problems have been created to provide opportunities for direct practice in sizing and resizing illustrations and photographs by both hand and computer methods.

Unit 26   CREATING GRAPHS AND CHARTS

BASIC PRINCIPLES OF CREATING GRAPHS AND CHARTS

Graphs and *charts* are frequently made by business and management personnel to help their employees, stockholders, and themselves understand the financial status of the entire company or any portion of the company. Before graphs and charts can be created, it is necessary to acquire numerical figures from which to build the pictorial image(s) that represent the numbers. For many, and possibly most people, a listing of numbers is difficult to interpret, but when graphs and charts are made, the numbers can be more easily understood.

Computer software is available that makes it rather easy to prepare bar graphs and pie charts. In many cases, the formats of the graphs and charts are already in *template form* and all that is necessary is to add the words, numbers, and sizes of the bars or pie shapes. Often colors are added to make the visuals more attractive and easy to understand. With full-color ink-jet and laser printers, colorful graphs and charts can be a reality with little effort.

PRACTICAL PROBLEMS

Note: The annual gross income of ABC QUICK PRINT from the previous year is shown in the bar graph in Figure 26–1. Respond to the questions in problems 1–8 based on the information shown in the bar graph.

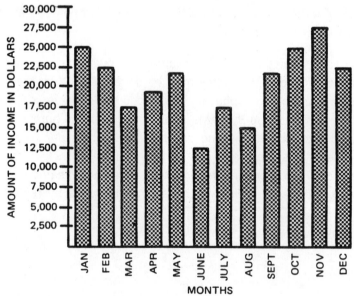

Figure 26–1   A bar graph in which the monthly income of a company is clearly illustrated.

1. What was the gross income for the month of January? _____

2. What was the gross income for the month of July? _____

3. What was the gross income for the month of December? _____

4. What was the gross income for the combined months of September and October? _____

5. What is the difference in income between the highest month and the lowest month? _____

6. What was the average monthly gross income for the first six months of the year? _____

7. What is the difference in the average monthly gross income for the first six months of the year and that for the last six months of the year? _____

8. What was the average monthly gross income for the entire twelve months of the year? _____

Note: Jennifer, a graphic arts company accounts representative, made some good sales during the past fiscal year. Draw a bar graph, either by the traditional t-square and triangle method or by using a graphs and charts software computer program, to reflect Jennifer's specified sales figures as listed in problems 9–20.

9.	January ---------------	$18,000	15. July ------------------	$20,000
10.	February -------------	$29,500	16. August --------------	$18,000
11.	March ----------------	$17,500	17. September ----------	$17,500
12.	April ------------------	$15,000	18. October -------------	$19,000
13.	May -------------------	$17,000	19. November -----------	$22,000
14.	June ------------------	$18,000	20. December -----------	$21,000

Note: The required hours for prepress, press, and postpress work for producing letterheads, catalogs, and brochures can easily be shown in bar graph form (Figure 26–2). This information is valuable to management personnel, as it helps them decide which products should be produced in their plant. Respond to the questions in problems 21–25 based on the information displayed in the bar graph.

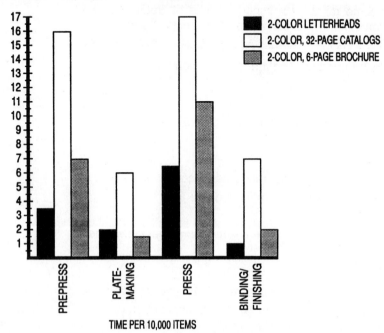

Figure 26–2   A multiple bar graph in which the hours of time required for different printed products in different phases of production are clearly illustrated.

21. How much time was required for the prepress phase of producing 10,000, two-color, 32-page catalogs?   _____

22. How much time was required to bind and finish 10,000, two-color, 32-page catalogs?   _____

23. How much time was required for the prepress phase of producing 10,000, two-color letterheads?   _____

24. How much total time did it take to produce the entire order for the 10,000, two-color letterheads?   _____

25. How much time was required for the platemaking and press phases for producing 10,000, two-color, 6-page brochures?   _____

Note: *Pie charts* are very useful for showing the division of costs, times, allotments, and other entities (Figure 26–3). In most cases, the entire circle is designed to represent 100% and the various divisions represent a portion of the total. For example, this pie chart is a representation of the $125,000 gross income of the Green River Stationery Printer for the month of October. Respond to the questions in problems 26–30 based on the information displayed in the pie chart.

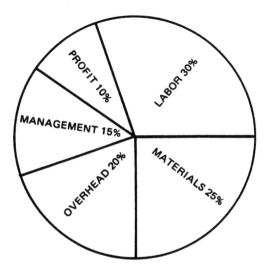

Figure 26–3   Pie charts, sometimes called circle graphs, are used to visually show how 100% of "anything" is divided into various categories.

26. How many of the gross income dollars were attributed to labor?   _____

27. How many of the gross income dollars were attributed to materials?   _____

28. How many of the gross income dollars were attributed to overhead?   _____

29. How many of the gross income dollars were attributed to management?   _____

30. How many of the gross income dollars were attributed to profit?   _____

Note: The Schultz Graphic Communications Company had a gross income of $255,000 for this past year. The president and owner of the company categorized the income dollars into the five areas of labor, materials, overhead, management, and profit. Create a pie chart representing the $255,000 and showing the information in percentage amounts as listed in problems 31–35.

31. Labor ----------- $102,000

32. Materials ----------- $76,500

33. Overhead ----------- $30,600

34. Management ----------- $24,225

35. Profit ----------- remainder of $

Note: *Line graphs* are very useful illustrations because they permit viewers to make some excellent comparisons between and among the various components (Figure 26–4). For example, the net production of printing presses can be easily compared on a line graph. Respond to the questions in problems 36–40 based on the information displayed in the line graph.

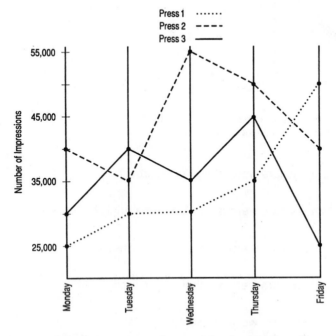

Figure 26–4 A multiple line graph can be used to visually show the productivity of people, machines, and processes.

36. How many impressions were printed on press 1 on Wednesday?   _____

37. How many impressions were printed on press 2 on Monday?   _____

38. How many impressions were printed on presses 1 and 3 on Thursday?   _____

39. What was the total number of impressions printed on press 2 for the entire five days of the week?   _____

40. What was the average number of impressions printed on all three presses for the five-day week?   _____

Unit 27   HAND-MECHANICAL SIZING METHODS

BASIC PRINCIPLES OF HAND-MECHANICAL SIZING METHODS

There are several methods that can be used to *resize* an image or measurement, but they are all based on the percentage of one measurement to another. For example, if a 2" × 4" rectangle is going to be enlarged, it is appropriate to enlarge both dimensions in the same proportion. **Note:** For standardization purposes, the first measurement given should always be considered as the width, thus the original rectangle is 2" wide. The second measurement should always be considered the vertical distance, thus the original rectangle is 4" high (Figure 27–1).

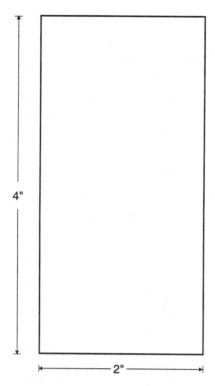

Figure 27–1   When giving and/or reading image or page measurements, the first measurement listed should always be considered the width and the second measurement should be considered the height. (Example: 2" × 4" rectangle; 2" = width, 4" = height.)

To enlarge this rectangle, or any other rectangle, 200% or twice the size, it is necessary to multiply each of the numbers by 200% or 2. The new rectangle size would be 4" × 8". This system works very well, but a faster method is the use of a *proportional wheel*, sometimes referred to as a *proportional dial* (Figure 27–2).

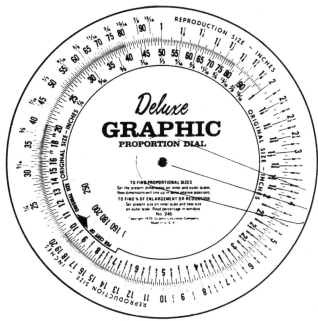

Figure 27–2 A proportional wheel, sometimes called a proportional dial, is very useful when determining reductions and enlargements of illustrations and photographs.

A proportional wheel includes two disks riveted together in the center so each disk can be turned independently of the other. The disks can be made of a paper board product or plastic sheeting. Exacting measurements are printed on both disks that permit the user to determine enlargements and reductions with ease. There is a "window" area showing the percentage of enlargement or reduction. It is important to know that 100% means same size, thus any number over 100% means an enlargement and any number below 100% means a reduction in size.

To use a proportional wheel, the original size number is found on the smaller of the two disks. That number is then aligned with the new size number (enlargement or reduction) on the larger disk. It is then possible to refer to the percentage window and read the percentage of enlargement or reduction. Again, this procedure can be accomplished using basic mathematics, but the proportional wheel makes the task very easy and accurate.

On some proportional wheels, an indicator arm has been fastened to the two disks with the center rivet. This indicator arm is not necessary, but it does make the task of aligning the selected measurements very simple. Also, once the original two measurements are aligned, the two disks can be held together and the indicator arm moved to determine new proportions. For example, a small photograph measuring 2⅛" × 3½" must be made into a halftone that is 1¼" wide, and the question is how high the reduced photograph will be. To determine the new measurement, it is necessary to align the original measurement of 2⅛" from the inner disk with the 1¼" measurement on the outer disk (Figure 27–3). When this has been accomplished, the indicator arm should then be moved to the 3½" measurement on the inner disk. To obtain the new reduction size for the photograph height, it is necessary to read the measurement found on the outer disk that is aligned with the 3½" measurement of the inner disk. In this example, the new measurement is 2¹⁄₁₆". Thus, the new size of the reduced photograph is 1¼" × 2¹⁄₁₆".

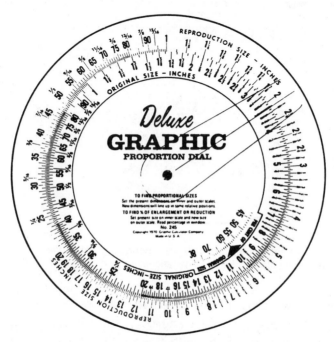

Figure 27–3   Proper alignment of the inner and outer disks of a proportional wheel is critical when determining enlargement and reduction measurements as well as the percentage of enlargement or reduction.

PRACTICAL PROBLEMS

1. Given the original size of 1" × 3¼", enlarge the width to 1½". What is the new height in inches and the percent of original size?

    _____ in

    _____ %

2. Given the original size of 2½" × 5¼", reduce the width to 1⅞". What is the new height in inches and the percent of original size?

    _____ in

    _____ %

3. Given the original size of 1" × 3", reduce the height to 1⅞". What is the new width in inches and the percent of original size?

_____ in

_____ %

4. Given the original size of ⅝" × 3", reduce the height to 1¹³⁄₁₆". What is the new width in inches and the percent of original size?

_____ in

_____ %

5. Given the original size of 2½" × 10", enlarge the width to 3". What is the new height in inches and the percent of original size?

_____ in

_____ %

6. Given the original size of 4" × 7", enlarge the width to 7⁵⁄₁₆". What is the new height in inches and the percent of original size?

_____ in

_____ %

7. After cropping, a photograph taken by the newspaper staff photographer measured 6" × 8". What will be the width of the halftone that will appear in the daily newspaper if the height is three inches?

_____

8. An artist brought a drawing she had prepared to a prepress specialty company and wanted a film negative made. Because the original drawing was 18" × 24", a traditional process camera had to be used because the flatbed scanner was not large enough to accommodate a drawing this large. The artist wanted the drawing reduced in size so the image depth would fit a space 11" in height. What will be the drawing width when the camera operator reduces the 24" to 11"?

_____

9. A photograph 5" × 6" is going to be used for a full-page advertisement that will be printed as an 8½" × 11" flyer. Which measurement of the original photograph will need to be cropped?

_____

10. A snapshot that measures 13 picas × 28 picas is to be scanned, processed via the appropriate software, and then made into a halftone using the imagesetter. The specifications call for the halftone to be 9 picas wide. How many picas high will the halftone image be?

_____

11. An illustration that was hand drawn by an amateur artist was selected to appear in a city chamber of commerce promotional booklet. The original drawing measured 50 picas × 75 picas. What percentage reduction is required for the drawing to appropriately fit into a space that is 32 picas × 48 picas?

    _____

12. The cropped area of a photograph measured 30 picas × 60 picas. What will be the height in inches of the photograph if the width is reduced to 2¾"?

    _____

13. An original drawing created by a local artist is 25 cm × 40 cm. What will the width be in centimeters if the height is reduced to 33 cm?

    _____

14. The photograph of the class president was recently taken, but there was an error when the print was made. The print should have been sized to fit a frame for a 20 cm × 25 cm photograph but was sized 18 cm × 22.5 cm instead. What percentage of enlargement increase is needed for the 18 cm × 22.5 cm photograph to fit the 20 cm × 25 cm frame?

    _____

15. A space was left in a column of the *Metric Newsletter* for a halftone print 5 cm × 8 cm. The photograph provided by the staff photographer was 13 cm × 20 cm in size. Which of the two measurements—width or height—had to be cropped so the other measurement would be large enough when the photograph was reduced in size?

    _____

Unit 28 COMPUTER SIZING METHODS

BASIC PRINCIPLES OF COMPUTER SIZING METHODS

Computers are but a tool for helping people accomplish one or more tasks at a given time. It must be quickly stated, though, that computers are very sophisticated tools and that without them, it would be difficult to accomplish many tasks that have become very common (Figure 28–1). For example, changing the size of an illustration or photograph once it has been included into a computer software file is convenient to accomplish. If the width of a photograph is in need of being changed, either larger or smaller, the height is automatically increased or reduced in proportion to the new width. The creators of the selected computer software have written that feature into the software program, thus the computer user does not need to be concerned about how much to increase or decrease the height when a new width is established. On the other hand, illustration and photograph sizes can be changed manually without regard to retaining the original proportions. Both of these functions are useful features of page creation software.

Figure 28–1 Sizing and resizing of illustrations and photographs can be easily accomplished using page creation software. (Credit: Apple Computer, Inc.)

PRACTICAL PROBLEMS

Note: The following problems are based on one or more of the common *page creation software* programs that are typically used in the graphic communications industry and by anyone else involved in creating pages containing text, illustrations, and photographs.

1.   A 2" × 3" photograph was scanned into a page software program. The graphic designer decided that it was necessary to increase the width of the photograph to 2.5". What was the new height of the photograph?   _____

2.   A customer brought the copy and several photographs to the Valley Quick Print graphic communications company. The purpose was for the designers and copy preparation personnel at Valley Quick Print to create a 6-panel flyer that would be printed on an 8½" × 11" sheet. To create the six panels or pages, the 8½" × 11" sheet had to be folded into a letterfold that consisted of two parallel folds equally dividing the 11" sheet and making each of the pages 3⅝" in width. One of the photographs was 4" × 6" with the content in a vertical format. The designer decided to leave ¼" margins on each panel and have all copy extend to the margins on both sides. Using page creation software, what was the reduced width of the photograph when the designer brought it into the page?   _____

3.   Utilizing the information in problem 2, it can be understood that the designer needed to know the new height of the reduced 4" × 6" photograph. Based on the reduced width of the photograph, what was the new height (reduced height) of the 4" × 6" photograph?   _____

4.   Utilizing the information in problem 2, the designer decided to create an illustration, using the page creation software, that would fill the entire page area of one panel or page of the flyer. It must be kept in mind that the designer had decided to utilize ¼" margins on all four sides of each panel of the flyer. What size was the image area of the finished illustration?   _____

5.   An original illustration was 8" × 5" in size. It was scanned into a page creation software by a graphic designer at the Cedar Communications Company. The 5" measurement had to be enlarged to fit a 12" × 9" allocated space on a planned poster. How much of the 8" dimension of the illustration had to be cropped to fit the 12" allocated space?   _____

6. A computer operator, using a page creation software, wanted to create an image block with straight lines of type. The area was to be 2⅛" inches square, thus the lines needed to be exactly 2⅛" long. The computer operator decided that the lines would be 1⁄16" wide and that there would be ⅛" between each line. How many lines were needed to fill the 2⅛" height? _____

7. Upon creating the square image area described in problem 6, the computer operator was interested to know how many white line spaces would result when the 1⁄16" width lines were added to the 2⅛" square area. How many white line spaces did result when the 2⅛" image block area was created with the 1⁄16" lines? _____

8. A logo for a nationwide insurance company was being included on a small calendar that could be attached to a computer monitor or the dashboard of an automobile. The original logo that was available to the designer was 3" × 2¾". The allocated width for the logo on the layout of the small calendar was 1 1⁄16". What was the height of the logo in inches after it was reduced in size by the computer operator? _____

9. Judith, a creative designer, was in process of designing a box cover using a popular page creation software. She used a photograph that was square and made it a rectangle that was in the proportion of 2 to 3. What percentage of the content had to be cropped from the width of the photograph? _____

10. An insurance identification card was being designed that would fit into a pocket of a standard men's and women's billfold, sometimes referred to as a wallet. The customer, a representative of an insurance company, requested that a solid area of color be included at the top of the card that would include the reverse images of the company logo and the name of the card. The card was to be 3⅜" × 2⅛" in size, and the solid area of color was to be ½" in height. How many times larger was the solid area of color wide than it was high? _____

11. The printed image that was going to appear on a computer mouse pad was being designed by a designer who worked in a screen printing company. He created a three-color layout, using a page creation software program, covering an area that was 8" × 6½" is size. There were three illustrations with dimensions of 1½" × 2", 4⅝" × 1⅜", and 8" × 1¾". What percentage of the total mouse pad image area was covered with illustrations? _____

12. The original negative of a 10" × 8" photograph was scanned into a page creation software program file. Upon inspecting the content of the photograph on the computer screen, the designer decided that a 1½" portion of the photograph could be cropped from the right side without detracting from the content of the photograph. By cropping this amount from the right side of the photograph, content was also cropped from the height of the photograph because the designer wanted the remaining content of the photograph to remain in the original proportion of 10" × 8". How much of the photograph height in inches had to be cropped to retain the same proportion of the original photograph?

                                                                          _____

13. Computer page creation software programs permit the creative designer to reduce or enlarge an illustration in most any proportion or format that is desired. Both horizontal and vertical crop lines can be positioned over an illustration that is being displayed on the computer screen. Using a 5" × 7" original photograph in a vertical format, the designer decided to make the final image of the photograph into a square format. With this action, what percentage of the original photograph was cropped out with the software crop line?

       _____

14. An electronic camera was being used to capture images of historical architecture for use in creating a promotional booklet for the chamber of commerce of a midwestern city. Once the images had been captured with magnetic storage in the camera, the images were downloaded into a page creation software for use when laying out the booklet. The proportional format of the electronic camera was the same as a standard 35mm camera designed for silver-based photographic film. Therefore, the images were in the proportion of 1⅜" × 1¹⁄₆₆". By doubling this actual size (an enlargement of 200%), how many photographs could be positioned within a vertical space of 4⅛"?

       _____

15. The graphic designer who was working on the chamber of commerce promotional booklet described in problem 14 was involved in determining an appropriate width for the promotional booklet. The designer created the boxes for three photographs, leaving ¼" between each box plus ½" margin on each side, that would appear in a horizontal format across the booklet (Figure 28–2). By enlarging each of the electronic images 163.5%, it was possible to fill each of the three boxes with photographic images. How wide

in inches did the promotional booklet have to be to accommodate the three photographs, the space between the photographs, and the margins?

_____

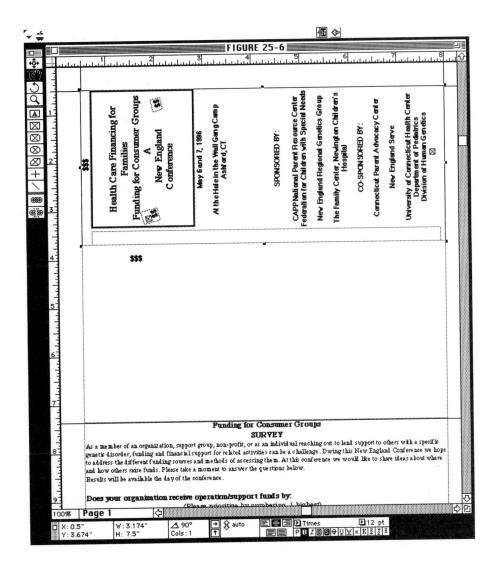

Figure 28–2   Image areas, space between the image areas, and margins can be created with page creation software prior to adding the actual photographic content. (Credit: Quark, Inc. and DK & A, Inc.© 1988–1994 Quark, Inc. All rights reserved.)

Computer Page Layout

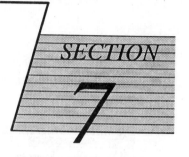

Computers have become a mainstay in graphic communications. There is hardly a phase of this technological area that has not been impacted by computer technology. Computers are sophisticated tools that make it possible to accomplish important tasks easier, faster, and better than what could previously be done by hand and other mechanical methods. Knowing how to utilize computer technology is essential for individuals to be successful in the graphic communications industry. This is also true for management personnel of graphic communications companies. They must provide both the *hardware* and *software* for use by their employees if they expect to be competitive and remain on the cutting edge of contemporary business and industry.

People need to know basic mathematics if they are to be successful in any aspect of the graphic communications industry. As stated previously, computers are essential tools for accomplishing a wide variety of tasks throughout all production and management phases of graphic communications. This is especially true in the copy preparation area. Knowing the fundamentals of determining spacing in type composition is as important today as it was yesterday. Letters, words, and paragraphs are the center of most printed and published works. The correct spacing of these three "visual" essentials is critical to the *readability* of the content.

Besides readability, *aesthetics* and overall visual appeal are important to the acceptance and understanding of printed pages. Margins must be carefully calculated and adjustments must be made in the computer software so the correct margin spacing is achieved. The correct positioning of image content in columns and tables is also essential so people will be willing to read the material and then be able to understand what they have read. For most applications associated with calculating margins and creating tabs, columns, and tables, the basic mathematical principles presented in Sections 1 through 5 are sufficient for being successful in preparing page layouts with computer hardware and software.

Unit 29   SPACING IN TYPE COMPOSITION

BASIC PRINCIPLES OF SPACING IN TYPE COMPOSITION

In order to make printed materials easier to read and to set off certain parts for emphasis purposes, space is usually left between lines of type. The typeset material you are reading at this very moment has spacing between the lines. If there were no line spacing, the letters in the lines would be so close together that it would be very difficult to read the type.

Spacing between lines of type is most frequently referred to as *leading*. This term is derived from the time of setting foundry type by hand that consisted of individual letters cast in metal. As is well known, Johann Gutenberg of Mainz, Germany, invented movable metal type in the middle 1440s. To separate the lines of type that were set by hand, thin strips of lead were placed between the lines. In part, this made it easier to handle typeforms made up of hundreds of pieces of individual type characters, but it also made the type much easier to read because of the added spacing. When mechanized typesetting came into existence in the 1880s, the term *leading* continued in use through the following decades. Computerized typesetting did not change or outdate the use of the term, thus it is as commonly used today as it was many years ago.

Leading, then, means the spacing between lines of type. Computer software programs written for page creation and special typesetting software have been written so the amount of leading between lines can be adjusted according to the wishes of the typographer or the person responsible for keyboarding the copy. Word processing software programs have not been written with adjustable leading; line spacing has already been determined according to the selected size of the type.

Too much leading causes the typeset page to look awkward while insufficient leading sometimes causes problems with lines of type appearing to overlap each other (Figure 29–1). As little as one-half point of leading may be used between lines of type; however, one or two points are more commonly used as leading amounts. People responsible for specifying type for a specific publication often use terminology such as, *set 8 on 10*. This specification also can be listed as "8/10." To "set 8 on 10" means that the type should be 8 point with 2 points of leading. The following are other examples: set 10 on 11 means one point of leading; set 10 on 12 means two points of leading; and set 12 on 13.5 means one and one-half points of leading. It is necessary to understand that the larger number means the total space consumed by the line of type and the space (leading) between the lines.

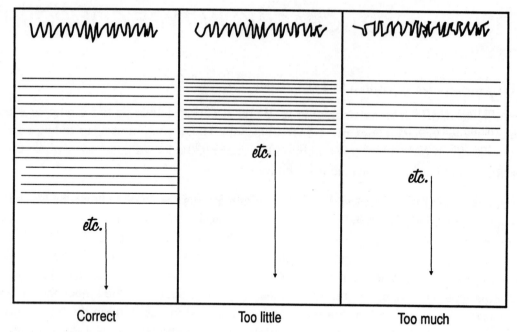

| Correct | Too little | Too much |

Figure 29–1 Comparisons of leading—correct, too little, and too much.

Determining the Pica Depth of a Page

When figuring the depth of any type area, count the number of lines and then subtract one line space or leading amount. For example, with 12 lines of type, there are 11 spaces between the lines. The problems in the following examples should help when calculating how much vertical page space will be consumed by given amounts of type lines and leading.

Example 1: Find the pica depth of a 24-line page of 8-point type with 2 points of leading.

Determine the point depth:

$(24 \times 8) + (23 \times 2) = 238$ points

Determine the pica depth:

$238 \div 12 = 19$ picas and 10 points

Example 2: What is the pica depth of a page of type set 9 on 12 and having 30 lines?

Determine the point depth:

$(30 \times 9) + (29 \times 3) = 357$ points

Determine the pica depth:

$357 \div 12 = 29$ picas and 9 points

Determining the Number of Type Lines in a Page

When figuring the number of lines in a given height or vertical area of type, the first step is to convert the inch or pica measurement to points. To convert inches to points, the measurement must be multiplied by 72 (there are approximately 72 points per inch), and to convert picas to points, the measurement must be multiplied by 12 (there are exactly 12 points per pica). Once the total points have been determined, divide by the type size plus the leading to find the lines of type per page. The problems in the following examples should help when calculating the number of type lines in a vertical page space.

Example 1: How many lines of 10-point type with 2 points of leading are contained in a page of type measuring 4" in height?

Determine the point depth:

$4 \times 72 = 288$ points

Determine the number of lines:

$288 \div (10 + 2) = 24$ lines per page

Check your work:

24 lines = $(24 \times 10) + (23 \times 2) = 286$ points

Since there are 288 points, there are 24 lines.

Example 2: How many lines of type are contained in a page 54 picas and 8 points long if the page is set in 6-point type and has 4 extra points of leading for each line?

Determine the point depth:

$(54 \times 12) + 8 = 656$ points

Determine the number of lines:

$656 \div (6 + 4) = 65.6$ lines

Check your work:

65 lines = $(65 \times 6) + (64 \times 4) = 646$ points

66 lines = $(66 \times 6) + (65 \times 4) = 656$ points

Since there are 656 points of page height, there are 66 lines.

PRACTICAL PROBLEMS

1. There are 20 lines of 9-point type with 2 points of leading. What is the point depth of this typeform?

   _____

2. There are 15 lines of 8-point type with 2 points of leading. What is the point depth of this typeform?

   _____

3. There are 30 lines of 10-point type with 1 point of leading in a page of type on the computer screen. What is the pica and point depth of this page?

   _____

4. There are 40 lines of 12-point type with 2 points of leading (12/14) in a page of type on the computer screen. What is the pica and point depth of this page?

   _____

5. There are 15 lines of 9.5 point type with 2 points of leading planned for an image block within a page on the computer screen. The image block is 20 picas in height. How much space in picas and points will still be available in the image block either above or below the 15 lines of type?

   _____

6. An advertising flyer is being prepared using a page creation software program. The final page size of the flyer will be 8½" × 11", and there will be ½" margins on all four sides. How many lines of 14/18-point type can be set in the vertical space of the flyer?

   _____

7. The copy for a small 11" × 8½" poster must be prepared. The designer decided that 30-point type with 6 points of leading would be used for the copy that was to cover only half of the vertical space on the poster sheet. How many lines of type will it be possible for the computer operator to set in the available vertical space?

   _____

8. Using the information in problem 7, the computer copy preparation person had to determine where to begin the first line of type on the page (Figure 29–2). To make the calculation, it was necessary for the computer operator to remember to use the points left over from the lines of type plus half of the vertical space that had already been specified by the designer. How far down from the top of the poster sheet in full picas should the top of the letters of the first line begin?

   _____

Figure 29–2 Computer copy preparation personnel must calculate the space available for lines of type as well as the space available within given margins. (Credit: Linotype-Hell Company.)

9. A package designer has established a space on a layout for some technical information that can be typeset in 8-point type. The planned space is to be 15 picas wide and 6¼" in height. How many lines of 8-point type with 2 points of leading can be set in the 6¼" high space? _____

10. For another job, the same package designer has created a space that is to be 17 picas wide and 48 picas high. Again, the designer is planning to use 8-point type with 2 points of leading. With these specifications, how many lines of type can be set in the allotted space? _____

11. For a weekly newspaper, the editor needed to know how long her editorial should be for the next edition. She was informed by the copy preparation person that there would be 10½" of column space available and that 10/12 point type would be used. The editor knew that she could count on 15 words per line, but she needed to know the number of lines of type that would fit in the 10½" column space. How many lines of type could the editor count on having available for her column? _____

12. The "fine print" of a legal document was being planned. It was decided by the legal advisor that 6-point type with 1 point of leading would be used and that a maximum of 24 picas of vertical space could be allotted for the content. Using these specifications, how many lines of type could be keyboarded into this available vertical space?

_____

13. The computer copy preparation person was busy preparing some original copy with one of the popular page creation software programs. He had decided to use 14-point type with 2.5 points of leading to fill a column that was to be 56 picas in height. He did, though, plan to use an illustration that would consume 1½" of vertical space. How many lines of 14/16.5 type could be set in the available space if the illustration were used?

_____

14. The computer copy preparation person responsible for preparing the original copy described in problem 13 decided to leave the illustration out because there was need for more text to fit into the allotted space. How many more lines of 14/16.5 type could be set in the 56-pica height when the 1½" illustration was removed?

_____

15. A graphic designer consulted a typography expert and requested assistance with selecting the type for a product booklet. The booklet layout included 30" of column space that could be filled with typeset material and two photographs. The typographer suggested that the graphic designer use 14/17-point type and photographs that were 3" and 4.5" in height. One photograph had to appear in the first ten column inches, and the second photograph had to be positioned at the top of the last page. How many lines of type could be included in the 30" column space along with the two photographs?

_____

Unit 30   *CALCULATING MARGINS*

BASIC PRINCIPLES OF CALCULATING MARGINS

Determining margins for printed products is important to the overall layout of the piece whether it be a single sheet or a multiple-page book. Margins help establish the overall appearance of a product that has been published and distributed widely. Sometimes people look at a printed piece and decide whether to read it or not to read it based solely on the size of the margins. Small margins give the appearance of the reading contents being rather traditional and heavy, whereas larger margins tend to give the feeling of lighter reading.

There are two basic methods of determining margins for an imaged page. One method involves using "part" amounts for each margin. The traditional *part method* involves the use of 3, 4, and 5 part amounts for each margin. Normally, the top margin includes three parts, the side margins include four parts, and the bottom margin includes five parts (Figure 30–1). Each part should represent a specific amount. If each part in the illustration represented ¼", the top margin would equal ¾", the side margins would equal 1" and the bottom margin would equal 1¼". Using this method, the size of the part can be any measurement that the designer chooses it to be.

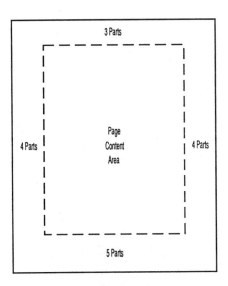

Figure 30–1 The "part" method of determining page margins involves placing three parts at the top margin, four parts at the side margins, and five parts at the bottom margin.

The second method involves the use of percentages to determine the size of the margin. The "percentage" method involves a greater understanding of margin sizes than the "part" method, but it is rather easy and fast to use. Typical percentages used to create visual vertical balance are 40% of the available space at the top, 50% of the available space on each side, and 60% at the bottom (Figure 30–2). This means that if the page is 6" wide and the image is 5" wide, each margin would be 50% of the remaining amount of one inch, thus each margin would be ½" in size. The same is true for the top and bottom except the percentage amounts of 40% and 60% are used instead.

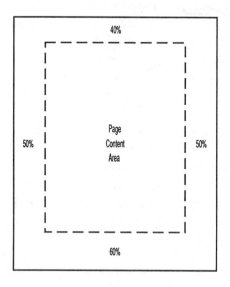

Figure 30–2 The "percentage" or "percent" method of determining page margins involves placing 40% of the total vertical margin space at the top margin and 60% at the bottom margin. The side margins are determined by taking 50% of the total vertical margin space for each of the two sides.

The entire purpose of the different "part" and "percentage" amounts is to create a condition called *visual balance*. Visual vertical balance simply means that there is more space in the bottom margin than there is in the top margin and that the side margins are generally a size in between the top and bottom margins. Equal margins, sometimes referred to as "mathematical balance," creates a condition that makes the imaged content look too low on the page. A well-designed book, a well-positioned photograph in a picture frame, and an accurately positioned pair of type columns on an advertising flyer will always have more margin space at the bottom than the top. In addition, the top, side, and bottom margins will be proportioned with either the "part" or "percentage" method.

With page creation software, margins can be easily established after the margin sizes have been determined. For the most part, margin calculations should be made prior to turning on the computer. It is a waste of computer time to "experiment" with margin sizes, thus work done beforehand is very valuable. The problems in this unit will provide you with opportunities to practice the basic mathematical skills developed in the earlier units of this textbook.

PRACTICAL PROBLEMS

1. A rectangular 3" × 6" illustration is to be positioned using vertical balance on a 7" × 10" sheet. What size in inches will the left and right margins be after the illustration has been positioned?

   _____

2. After the illustration was positioned left-to-right in problem 1, the computer operator had to position the illustration vertically. He had been informed that the illustration should be positioned by using the "part" method of establishing visual vertical balance. What size in inches will the top margin be using this guideline when the illustration is 6" high and the page is 10" in height?

   _____

3. The finished printed page will be the standard "business world" sheet size of 8½" × 11", and there will be two columns of typeset material. The lines of type will be 20 picas long, and there is to be 2 picas between the columns. The left and right margins must be of equal pica size. What size in picas will the left and right margins be on this sheet?

   _____

4. The computer copy preparation operator decided to use the "part" method of determining margins prior to keyboarding the two columns of type described in problem 3. She planned to use one pica as the measurement for each part, and as it turned out, the columns were 50 picas long. How large in picas will the bottom margin be?

   _____

5. A design and layout artist created the layout for a flyer using the "part" method of determining margins. Each page of the six-page flyer is to have three-pica margins on the left and right sides. What is the combined total of the top and bottom margins for each page?

   _____

6. A new high school graduation certificate was being designed by the school principal. Using the "part" method, she decided that the smallest margin would be two inches. What size in inches would the opposite margin need to be to meet the "part" method procedure?

   _____

7. The school principal introduced in problem 6 also needed to determine the size of the side margins. Using the standard "part" method of determining margins, what is the size in inches of the side margins for the graduation certificate?

   _____

8. The high school secretary was given the task of creating the new high school graduation certificate that the principal worked on as presented in problems 6 and 7. She was informed that the margin sizes determined in problems 6 and 7 must be used. In addition, she was given the certificate content to place into the remaining available space on the 11" × 8½" certificate (Figure 30–3). Prior to sitting down at her computer and using the page creation software, the secretary needed to determine the space available for the certificate content. Using the established margins determined in problems 6 and 7, what is the width in inches of the available content space?

_____

Figure 30–3   Determining margins and image content area are critical steps to complete prior to making computer software measurement settings.

9. After determining the width of the available content space for the high school graduation certificate as presented in problem 8, the secretary had to determine the amount of vertical space. How much vertical space in inches was available for the certificate content based on the determined top and bottom margins of the 11" × 8½" certificate?

_____

10. After beginning to keyboard some copy for a farm sale poster, the computer copy preparation specialist realized that the design and layout artist had not provided the suggested margins. He knew it was important to stop and make these determinations before advancing any farther with the job. He decided to use the 40, 50, and 60 percent ratios for establishing the margins. The poster was to be 11" × 14", which is considered a ¼ poster size. He arbitrarily decided that the side margins would be 10 picas wide. What would be the size of the bottom margin in picas, using the percent method of determining the four margins?

      _____

11. Obviously, the computer copy preparation specialist responsible for creating the farm sale poster presented in problem 10 needed to determine all four margins. Based on the side margins, what size in picas should the top margin be according to the calculations when determined with the percent method?

      _____

12. The graphic communications students, with the help of their advisor, decided to design, write, produce, and distribute a newsletter that they called *Graphically Speaking*. The newsletter was to be eight pages long, and each page was to be 8½" × 11" in size. The first page was to have two columns with 2 picas between the columns. The students decided that the side margins would be two and one-half times larger than the distance between the columns. What size in picas was each side margin of the newsletter page?

      _____

13. Continuing with the newsletter called *Graphically Speaking*, presented in problem 12, the students needed to determine the top and bottom margins by using the percent method of calculating margins. What was the combined top and bottom margin space?

      _____

14. After turning on the computer and entering the page creation software, the production coordinator of *Graphically Speaking* set the margins based on the information and calculations in problems 12 and 13. How wide in picas was each of the two columns of type based on the established margins and the space between the columns?

      _____

15. Using the percent method of determining margins, a book designer wanted to know the percent of margin space as compared to the content space of each page of a photography book. The book had page sizes of 8" × 9⅛", and the average content per page was 6¼" × 7¼" in size. What was the percent of the margin space on the four sides of each page as compared to the content of the overall 8" × 9⅛" book page?

_____

Unit 31 CREATING TABS, COLUMNS, AND TABLES

BASIC PRINCIPLES OF CREATING TABS, COLUMNS, AND TABLES

Computer users who work with page creation software must be well versed in creating tabs, columns, and tables. There are many kinds of forms where these three page formats are used. It is easy to find printed material in newspapers, magazines, company annual reports, advertising literature, and books where tabs, columns, and tables have been used to present graphic information in one or more of these formats. The procedure for determining where to create tabs, position columns, and prepare tables is directly aligned with basic mathematical calculations. Usually, it involves totaling the space needed for the planned or actual contents, finding the marginal areas and distances between the content, and dividing up the space. This sounds easy, but it can be confusing unless thumbnail sketches and/or rough layouts are prepared in advance of sitting down at the computer (Figure 31–1). Preplanning is critical! Preparing a sketch from the basic specifications can make the work of determining column sizes, tab stops, and margins much easier and more accurate. Experimenting at the keyboard is time-consuming and expensive.

Thumbnail Sketches

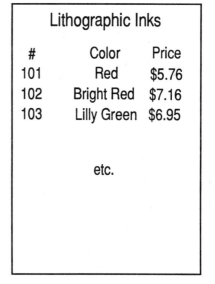

Rough Layout

Figure 31–1 Thumbnail sketches and rough layouts are very helpful when creating tabs, columns, and tables.

PRACTICAL PROBLEMS

1. The computer operator was asked to position three 1" high vertical lines across an 8½" wide page. The distances between the lines and the margins on both sides of the two outside lines had to be exactly the same. At what inch measurement was the tab for the first line (left) positioned?

   _____

2. After the tab was established for the first vertical line as presented in problem 1, the computer operator went ahead and established the second and third vertical line tabs. At what inch measurement was the tab for the third line (right) positioned?

   _____

3. Four capital letters were to be positioned equally across a page that was 10" wide. Each of the capital letters was 50 points wide. The specifications called for the distance between the four letters and the left and right margins to be exactly the same. The page creation software being used permitted the operator to set the tabs at specific point measurements. At what point measurement was the second tab set so the spacing would be as precise as the specifications?

   _____

4. Obviously, the computer operator had to set all tabs according to the information presented in problem 3. At what point measurement was the tab set for the fourth letter?

   _____

5. Six vertical lines, 5" high, were to be placed in specific locations on a 9" inch wide sheet of transparent plastic material. The first line was to be 1" from the left margin, the second line 2" from the left margin, and the third line 3" from the left margin. Lines four, five, and six were then to be positioned equal-distant between line three and the right edge of the sheet. At what tab setting did line five need to be established to meet the specifications?

   _____

6. The computer copy preparation specialist working on an eight-page newsletter noted from the rough layout that there were two columns of type, illustrations, and photographs planned for one of the 8½" × 11" newsletter pages. From the layout, he determined that the total space of the two margins was 7 picas and the specified distance between the two columns was 2 picas. The computer operator had to determine the line length for the two columns. What was the length in picas of the column lines for each of the two columns?

   _____

7. For several pages of the eight-page newsletter described in problem 6, there were to be three columns instead of two (Figure 31–2). According to the rough layout, the left and right margin specifications remained the same, but the distance between the columns was changed from 2 picas to 1 pica. What was the length in picas of the column lines for each of the three columns?

_____

Figure 31–2 The rough layout of a newsletter page in which there were to be three columns of type, illustrations, and photographs. (Credit: Quark, Inc. and DK & A, Inc. © 1988–1994 Quark, Inc. All rights reserved.)

8. On the back page of section one of a Pulitzer Prize winning newspaper, there were six columns of typewritten content. The usual 1-pica space was used between the columns, and there were two picas of margin on both the left and right sides of the page measuring 13½" wide. What was the line length in picas of each of the six columns of type?

    _____

9. Newspaper editors often utilize different column widths on the same page. This is done to add variety and provide sufficient space for selected content. On the same back page of the newspaper described in problem 8, there was a portion of the page that contained five columns instead of six columns. Interestingly enough, four of the columns were the same width as determined in problem 8, but the one remaining column was considerably wider. How wide in picas was this fifth column?

    _____

10. In yet another portion of the same newspaper page described in problems 8 and 9, there were only four columns. One column was the same width as that determined in problem 8 and one column was the same width as that determined in problem 9. What was the individual width in picas of the two remaining columns?

    _____

11. A table of numbers from 501 through 700 had to be created. There were to be twenty lines with ½ pica between each of the columns. How many columns of numbers had to be formed to accommodate the total table of numbers?

    _____

12. A table had to be designed that would appear in a textbook containing a content page width of 26 picas. There was to be a total of thirteen columns in the table. The left column was to be 10 picas wide, and the next eleven columns were to be the same size. If the right column were 5 picas wide, how wide in picas would each of the eleven columns in the center portion of the table be?

    _____

13. Another textbook table had to be prepared from an author sketch, but in this case, the table was horizontal instead of vertical. With this table, there were to be six horizontal columns the same size and one horizontal column larger to accommodate several lines of type. The book page width was 26 picas, as in problem 12. If each of the horizontal columns were 3.5 picas in height, what size in picas would the remaining larger horizontal column be?

    _____

14. The computer copy preparation specialist was challenged with the next table that had to be planned and composed. There had to be three columns, but each of the columns had to be a different size because the content in each column was of a different nature. In fact, column one contained paragraphs, column two contained product names, and column three contained six numbers such as 25,683. The table was to be 41 picas wide to correspond with the width of the multiple columns throughout the book. What was the total number of picas of content width in the three columns, knowing that the spacing between the columns was 3 picas each?   _____

15. The final table in the photography textbook had to contain six columns to accommodate the required information. One column had to be 6 picas wide, and the other five columns had to be 3 picas wide. As stated in problem 14, the content page of the textbook was 41 picas wide. How much space in picas had to be left between each of the columns to make the table look balanced?   _____

Paper Stock Needs

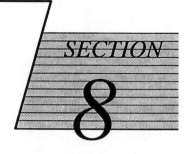

Paper is the primary medium or *substrate* used for imaging purposes in the graphic communications **industry.** Some reasons for this fact is that paper readily accepts images, it is economical, it is light in **weight,** it can be folded into many sizes and shapes, and it is a renewable product. With the majority of paper being made from trees, it is possible to manufacture nearly any amount of paper needed to meet the demand. This is true because trees can be and are grown as recycled plants. When a forest of trees is cut for the purpose of making paper, new trees are immediately planted to replace them. It generally takes ten to fifteen years for trees to reach the appropriate *pulpwood* size, but this is considered a short time in the overall life cycle of most trees that are common to the countryside.

There are other substrates used for printing and imaging. These include glass, cloth fabric, wood, **metal, and the** many plastics that have been created over the years. Special knowledge and handling **are required** for using these materials, but beautifully imaged products from these materials are all **around us.**

There is much to know about paper because there are many kinds, colors, sizes, and finishes of paper. Knowing how to utilize mathematics in calculating paper sizes, thicknesses, and weights is essential to being successful in this phase of the graphic communications industry. Also, anyone involved in job planning must know how to calculate the number of *press sheets* from *stock sheets*. Finally, people **with the** responsibility of determining the amount of paper required for printed jobs must possess **sufficient** mathematical abilities to make the correct calculations. Along with the required basic **mathematics,** these people must possess "thinking" skills so they can accurately determine paper **needs.** The basic principles information and problems in each of the three units in this section have **been written** to improve your thinking and mathematical skills.

Unit 32 BASIC SIZE, THICKNESS, AND WEIGHT OF PAPER STOCK

BASIC PRINCIPLES OF SIZE, THICKNESS, AND WEIGHT OF PAPER STOCK

There is much to learn and know about writing and printing papers. In fact, it takes people who work full-time in the paper industry many years to learn everything there is to know about papers. Of course, paper technology is continually changing, thus for people who make their living in the paper industry, there is always something new to learn. The following information and problems about writing and printing papers are a sampling of what there is to learn and know about papers. Master this information and the subsequent problems, and you will be on your way to becoming an "expert" about writing and printing papers.

Paper is manufactured in a variety of sizes, weights, colors, and compositions. It may be ordered in many different sizes, but all paper is classified into basic sizes (Figure 32–1). For the most part, the

Bond and Ledgers
    Writing, Copier,
    Duplicator, Gummed, Flats
	17 × 22	Carbon papers
		    Pencil, Typewriter, Copying
			22 × 34

Blotting Papers
    Plain, Coated
	19 × 24	Card Stock
		    Rope, Bogus, Tag, Folding,
		    Postcard Ticket, Wedding
			24 × 36

Blanks
    Translucents, Railroad,
    Tough check, Poster,
    Calendar, Sign Board,
    Playing Card
	22 × 28	Cover papers
		    Specialty covers
			20 × 26

		Manilas Poster, Kraft	24 × 36

Boards
    Binders, Cloth Corrugated
	20 × 30	Newsprint
		    Poster paper
			24 × 36

Book papers
    Offset, Label, Text, Ballot,
    Coated, Uncoated
	25 × 38	Padding board
		    Strawboard, Chipboard,
		    Boxboard
			26 × 38

Bristols
    Mill
 	
    Index
	25½ × 30½	Pressboard
		    Tagboard
			24 × 36

		Thin papers
		    Manifold, Onion skin,
		    Tissues, Parchment
			17 × 22

Figure 32–1 The basic categories, subcategories, and sizes of writing and printing papers.

basic sizes of each paper category were determined many years ago according to the typical uses of the paper. For example, bond papers are most commonly used for letterhead, forms, and information sheets relating to business functions. The standard size for business letterhead and other business-related sheets is 8½" × 11". This is true in the United States, but other basic business sizes are used in other countries of the world. With the basic size of bond papers being 17" × 22", it is possible to obtain four 8½" × 11" sheets from this one sheet without having waste (see Unit 31).

Paper thickness is commonly measured in *thousandths of an inch*. For thicker papers, sometimes referred to as boards, the thickness is stated in *points*. A point, when referenced to paper thickness, is equal to $\frac{1}{1000}$ inch (.001 inch). This "paper" point has no relationship to the linear point presented in Unit 20 of this textbook. Another term often used with paper thickness is *caliper*. Paper thickness and caliper mean the same thing, which is the thickness of paper measured in thousandths of an inch. It is valuable to know the caliper of the various papers because this helps when designing a printed product and selecting the best paper to use (Figure 32–2). The wrong paper choice may tend to bulk up a product such as a book more than is appropriate. The problems in the following examples should help in determining the thickness of finished products and the papers used to manufacture the finished products.

Bond
Basis 17 × 22
20/40M 004

Ledger
Basis 17 × 22
28/56M 005

Manifold
Basis 17 × 22
9/18M 0015

Book Basis 25 × 38
Coated Book
60/120M 003

Eggshell Book
60/120M 006

English Finish Book
60/120M 004

Uncoated Book
60/120M 0045

Cover
Basis 20 × 26
Plain Cover
65/130M 0095

Coated Cover
60/120M 0055

Bristol
Index Bristol
Basis 25½ × 30½
110/220M 00825

Printing Bristols
Basis 22½ × 28½
100/200M Plate 0085
100/200M Antique 011

Blank, Railroad Board
4 ply 018

Tagboard
Basis 24 × 36
100/200M 008

This chart contains a token amount of the entries found in charts in paper catalogs.

Figure 32–2 The basic categories, sizes, weight per ream per basic size, and the basic caliper or thickness of papers in thousandths of an inch.

Example 1: A book contains 450 pages and each sheet of paper is .0035-inch thick. How thick in inches is the total book?

450 pages (2 pages per sheet) = 225 sheets of paper

225 × .0035 = .7875 inch thick

Example 2: Four sheets of book paper are 16 thousandths of an inch thick. How thick in inches are 248 pages?

One sheet of book paper = .016 ÷ 4 = .004 inch

248 pages = 124 sheets of paper

124 × .004 = .496 inch thick

The *basis weight* of paper refers to how much a ream of paper weighs when cut to its basic size. For example, 500 sheets of designated 20-pound bond paper, size 17" × 22", will actually weigh 20 pounds when placed on a scale. Paper weights are listed in several ways, thus it can be confusing. Twenty-pound bond paper that is 17" × 22" can be listed in six different ways on the label of a paper package or in a price book:

17 × 22 - 20	meaning 20 pounds per 500 sheets, per basic size
17 × 22 - Sub. 20	meaning 20 pounds per 500 sheets, per basic size
17 × 22 - 20D	meaning 20 pounds per 500 sheets, per basic size
17 × 22 - 40M (S20)	meaning 40 pounds per 1,000 sheets, per basic size
17 × 22 - 40M	meaning 40 pounds per 1,000 sheets, per basic size
17 × 22 - Basis 20	meaning 20 pounds per 500 sheets, per basic size

The direction of the *grain of paper* is important to both the press and finishing operations of a printed product. In short, knowing the paper grain makes it much easier to run paper through the printing press and through several of the finishing and binding operations. For example, paper bends and folds easier with the grain than against the grain. Grain direction can be determined several ways, but the easiest method is to look on the label of the paper package (Figure 32–3). The direction of the grain is either underlined in the stated dimensions (for example, 17" × 22") or marked "short" or "long" grain on the label.

17 × <u>22</u>–16 Substance 16 D16 lb M32 lb	Name Brand BOND PAPER	WHITE One Ream 500 Sheets

Figure 32–3 A paper package label that contains useful information for the user. The meaning of the information in the left-side square: 17 × 22 = size and grain direction; –16 = weight per basic size per 500 sheets; Substance 16 = basis weight; D16 lb = basis weight per 500 sheets; and M32 lb = weight of basic size per 1,000 sheets.

PRACTICAL PROBLEMS

1. Bond papers are frequently used in business operations for products and services such as letterhead, forms, and photocopiers. What is the basic size of bond papers?   _____

2. The book category of papers was created for use in the manufacture of books, whether they be textbooks, novels, books for children, or research publications. What is the basic size of book papers?   _____

3. Cover papers were originally produced to serve as covers for business reports, small booklets, and professional association report covers. This is still the common use of cover papers, although they can be and are used for many printed and imaged products. What is the basic size of cover papers?   _____

4. Index-bristol papers are used for producing cards such as postcards, file cards, and notebook divider sheets. What is the basic size of index-bristol papers?   _____

5. Most paper is measured in thousandths of an inch. What is the typical thickness of 110-pound index-bristol paper in thousandths of an inch?   _____

6. The terms *caliper* and *thickness* essentially have the same meaning. What is the typical caliper of 20-pound bond paper?   _____

7. A 256-page book, exclusive of the covers, was printed with a soft, bulky paper that had a caliper of .006". How thick in inches was the body of the book?   _____

8. Six sheets of a 60-pound book paper have a caliper of .024". How thick in inches are 246 sheets?   _____

9. A 750-sheet stack of 60-pound book paper, often referred to as a *lift of paper*, measured 3" thick. How thick was each 60-pound sheet?   _____

10. Four sheets of 65-pound cover paper measured .038" thick. How thick in inches are 248 pages?   _____

11. The designer of a hardcover book was informed by the customer that the thickness of the 540-page book could not exceed 1⅝". What is the maximum caliper of the paper that can be used for printing the book?   _____

12. There were 850 sheets of 17" × 22" 20-pound bond paper in the storeroom. How much did the 850 sheets weigh?   _____

13. Upon conducting the inventory of the paper supply, the estimator found eight 100-sheet packages of 90-pound index-bristol paper that were cut to the basic size for this paper category. He also found fifteen reams, basic size, of 50-pound book paper. What was the total weight of the paper the estimator found in the paper supply?   _____

14. Paper package labels contain considerable information (Figure 32–4). How many square inches of paper are contained in the one package of paper identified in Figure 32–4?   _____

BRAND NAME	25½ x 30½-220M	B 110
	648 x 775 mm	a 199 g/m²
INDEX	100 Sheets GRAIN LONG	g
	ITEM CODE	s
	23097	GREEN

Figure 32–4   Considerable information can be extracted from a paper package label.

15. Paper is not always packaged in ream amounts as shown on some package labels (Figure 32–5). How many square feet of paper are contained in one-half of the sheets in the package shown in Figure 32–5?   _____ _____

                         BRAND Opaque Cover Vellum

NAME   <u>23</u> × 35–100 1/2   201M Bs. 65

ADDRESS   584 × 889 mm   176 g/m^2

Citrus Yellow   750 Shts 455-1401

Figure 32–5 It is important to carefully read each paper package label so correct information is acquired.

Unit 33 DETERMINING PRESS SHEETS FROM STOCK SHEETS

BASIC PRINCIPLES OF DETERMINING PRESS SHEETS FROM STOCK SHEETS

Graphic designers and estimators should always know the basic sizes and standard sizes of the paper they are planning to use for a given product. It would be unwise to plan a finished printed product of a size that would not cut economically out of a standard stock sheet of paper. With the basic size of bond papers being 17" × 22", it can easily be seen that 8½" × 11" letterheads cut out perfectly, thus there is no waste (Figure 33–1). This does not always work out as easily as just illustrated, but the objective should always relate to obtaining the "best" cut while still meeting the production and customer requirements such as grain direction.

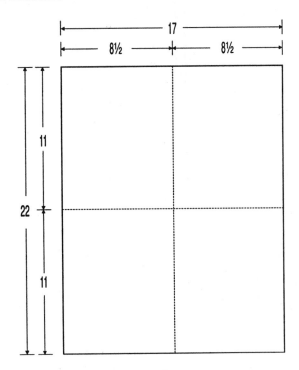

Figure 33–1   Standard letterheads, 8½" × 11", cut perfectly from bond paper that is of the basic size, 17" × 22".

The designer of the printed product must decide if the paper needed for the job should be cut with the grain, across the grain, or if the direction of the grain does not matter. The grain of paper is important when producing the product, and it is often very important for the final use of a product. As is well known, paper folds easier with the grain. Books open with ease and stay open when the grain runs vertical with the page. Posters printed on poster board material stand without sagging when the grain is vertical. On the other hand, it is best to run heavier papers and boards through sheet-fed printing presses with the grain parallel to the cylinders so the paper or board material will easily bend while it is being printed and handled. Study Figures 33–2 through 33–5 to learn how 5" × 3" sheets can be cut from a 17" × 22" sheet and retain the correct grain direction. Remember that a line under a measurement indicates the grain direction of the paper.

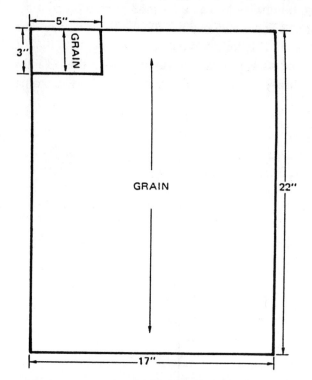

Figure 33–2   How many pieces 5" × 3" can be cut from a sheet of paper stock that is 17" × 22" if the grain must be with the 3" measurement?

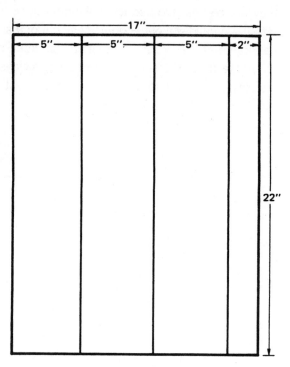

Figure 33–3   Step one is to divide the width of the paper stock by the width of the cut piece. The result: 17 ÷ 5 = 3 with 2" waste along the right side.

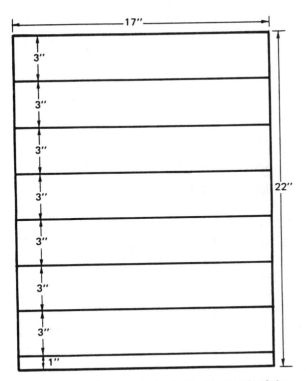

Figure 33–4 Step two is to divide the length of the stock paper by the length of the cut piece. The result: 22 ÷ 3 = 7 with 1" waste.

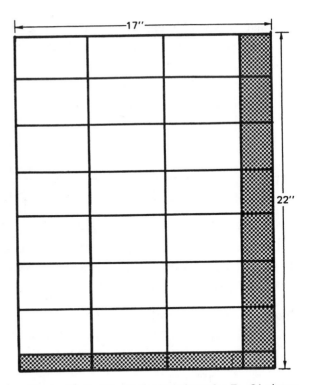

Figure 33–5 The final calculations: 3 × 7 = 21 pieces and 15.78% waste.

It is also important to determine the amount of waste that occurs when smaller sheets, often referred to as *press sheets*, are cut from *stock sheets*. The waste shown in Figure 33–5 amounts to 15.78%. It is calculated as follows:

Stock sheet:	17" × 22"
Possible square inches:	17" × 22" = 374
Used square inches:	(3" × 5") × 21 = 315
Waste:	374 sq in − 315 sq in = 59
Percent of waste:	59 sq in ÷ 374 sq in = 15.78%

When the direction of the grain is not an important factor, the designer or estimator determines the best cut and chooses the way in which the most pieces will be obtained. In this procedure, fractional pieces of paper become waste that reflects in the overall cost of the job. The following two examples provide insight into how small sheets (press sheets) can be calculated from stock sheets.

Example 1: How many pieces 5" × 3" can be cut from a sheet of stock 19" × 25"? Neither grain direction nor the amount of waste should be considered in this solution. The objective is to obtain the greatest number of 5" × 3" sheets.

$$\frac{19" \times 25"}{5" \times 3"} \quad \text{meaning} \quad \frac{19" \div 5" = 3}{25" \div 3" = 8} \quad \text{thus } 3 \times 8 = 24 \text{ pieces}$$

3 × 8 = 24 pieces

or

$$\begin{array}{c} 19" \times 25" \\ \times \\ 5" \times 3" \end{array} \quad \text{meaning} \quad \frac{19" \div 3" = 6}{25" \div 5" = 5} \quad \text{thus } 6 \times 5 = 30 \text{ pieces}$$

5 × 6 = 30 pieces

The most economical cut would be the latter, 30 pieces.

Example 2: How many pieces 5" × 7" can be cut from a sheet of stock 23" × 29"? Again, grain direction and waste are not to be considered. Simply obtain the greatest number of small (press) sheets.

$$\frac{23" \times 29"}{5" \times 7"} \quad \text{meaning} \quad \frac{23" \div 5" = 4}{29" \div 7" = 4} \quad \text{thus } 4 \times 4 = 16 \text{ pieces}$$

4 × 4 = 16 pieces

or

23" × 29"

×   meaning $\dfrac{23" \div 7" = 3}{29" \div 5" = 5}$ thus 3 × 5 = 15 pieces

5" × 7"

5 × 3 = 15 pieces

The former, 16 pieces, would be the most economical cut.

In some situations, it is advisable to cut smaller (press) sheets from stock sheets by using the normal waste strip on either side of the stock sheet. In cases like these, it is often possible to gain from two to six extra small (press) sheets. There are times when grain direction does not make any difference either in the printing or the use of the product. Making sketches of the planned cutting procedure is perfectly normal and highly recommended because the cuts can become rather confusing. Study Figures 33–6 through 33–9 to learn how additional 3" × 5" sheets can be obtained from a 23" × 29" stock sheet when the possible waste is utilized. When this method is used, the percentage of waste is reduced from 21.29% to 12.29%.

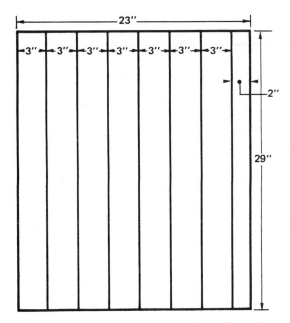

Figure 33–6 How many pieces 3" × 5" without regard to grain can be cut from a sheet of paper stock 23" × 29"? Step one is to divide the 23" width of the stock sheet by the 3" length of the small piece. The result: 23 ÷ 3 = 7 with 2" waste.

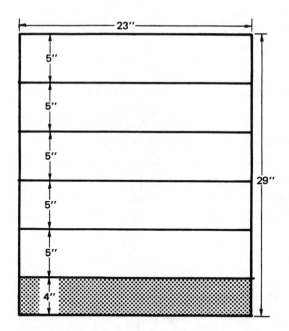

Figure 33–7   Step two is to divide the 29" length of the stock sheet by the 5" length of the small piece. The result: 29 ÷ 5 = 5 with 4" waste.

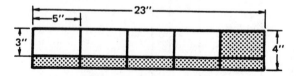

Figure 33–8   Step three is to determine the number of 3" × 5" small pieces that can be obtained from the 4" × 23" strip. The result: 4 ÷ 3 = 1 with 1" waste and 23 ÷ 5 = 4 with 3" waste.

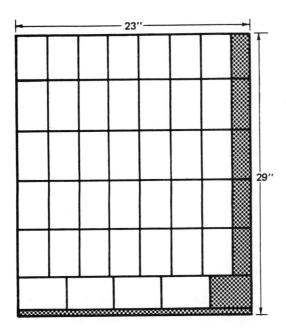

Figure 33–9   The final result: 7 × 5 = 35 + (1 × 4) = 39 pieces and 12.29% waste.

PRACTICAL PROBLEMS

1. It is necessary to cut 5" × <u>3</u>" small (press) sheets from stock sheets measuring 22" × <u>34</u>" and to observe the grain direction. How many small sheets can be cut from one stock sheet?

2. The binding and finishing supervisor asked Carl to cut some 6" × <u>9</u>" sheets from stock sheets measuring 17½" × <u>22½</u>" and to take care in observing the grain direction. How many 6" × 9" sheets can be cut from one 17½" × <u>22½</u>" stock sheet?

3. In addition, Carl was asked to determine how many <u>4</u>" × 6" scratch pad–sized sheets could be cut from stock sheets measuring <u>17</u>" × 22". He was asked to carefully observe the correct grain direction. What answer did Carl provide to his binding and finishing supervisor?

4. A 6" × 9" book was in process of being planned, and the designer needed to know how many book pages could be cut from a stock sheet measuring 25" × 38". Because the grain direction of the book pages was important to observe, the designer was careful to note the grain direction of the stock sheets when he calculated the needed information. How many book-sized pages could be cut from the selected stock sheets while retaining the correct grain direction?

_____

5. A special-sized notesheet holder, made of a bright-colored plastic, was being designed for sale in an office supply store. To determine the production cost of the notesheet holder, the designer needed to know the number of 2½" × 4¾" notesheets that could be cut from bond paper measuring 17½" × 22½". So the paper would lay evenly in the holder, the designer closely observed the grain direction. How many notesheets can be obtained from the stock bond paper?

_____

6. It was necessary to cut some 4" × 6" scratch sheets from stock sheets measuring 25" × 38". What is the percentage of waste when this size of scratch sheet is cut from this large stock sheet?

_____

7. Some file divider sheets needed to be cut for use by Mary, the office secretary. She stated that they needed to be 12¾" × 9½" in size. The sales representative of the local paper supply company stated that he had some cardboard stock available that was 36" × 48" in size. What is the percentage of waste based on the divider sheet size and the stock sheet size?

_____

8. Graphic communications people know that the basic size of bond paper is 17" × 22". Unless some mathematical calculations are done, though, most of these people do not know the percentage of waste when 5" × 7" notesheets are cut from the basic size. By closely observing the identified grain direction, what percentage of waste is incurred when this size notesheet is cut from bond paper of the basic size?

_____

9. To make sure the best cut was being made, the person responsible for the 5" × 7" notesheets decided to determine the percent waste if the grain direction was changed. The basic size of bond paper (17" × 22") was to be used, as stated in problem 8. What is the difference in the percentage of waste when the 5" × 7" notesheet grain direction is changed to 5" × 7"?

_____

10. Some 8½" × <u>11</u>" notebook sheets were needed by Cheryl, a graphic communications trainer, for a class she was going to teach. To help in determining the cost of assembling the notebooks, she calculated the number of notebook sheets that could be obtained from stock sheets of book paper that were <u>25</u>" × 38". Because cost was critical, she needed to know the percentage of waste while still closely observing the grain direction of the finished notebook sheets. The maximum amount of waste that she could accept was 20%. Was Cheryl able to use the <u>25</u>" × 38" stock paper based on her maximum waste guideline?   _____

11. Some 3" × 5" cards were to be cut from stock sheets measuring 17" × <u>28</u>". Not considering the grain direction, what is the maximum number of cards that could be cut from one stock sheet?   _____

12. The designer identified in problem 5 continued coming up with ideas for different sizes and shapes of colorful plastic notesheet holders. This time the designer was experimenting with notesheets that were 2½" × 5" in size. The available stock paper was <u>24</u>" × 38", and he did not care about the grain direction. What was the maximum number of 2½" × 5" notesheets that the designer could obtain from one sheet of <u>24</u>" × 38" stock paper?   _____

13. Some <u>4⅜</u>" × 6" index cards were needed as dividers for several hundred 4" × 6" photographs that had been taken by a group of European travelers. The travel tour guide requested that these be obtained at a nearby quick-print company. The company manager did some calculations according to the basic 25½" × <u>30½</u>" stock sheet size of 110-pound index paper. It was important to observe the correct grain direction so the cards would stand vertically in the photograph file drawers. How many file cards could be obtained from one stock sheet of index paper?   _____

14. The binding and finishing supervisor found some paper (20-pound bond) he had forgotten about and, because it was rather old, decided to simply cut it up for scratch paper. The bond paper was <u>22</u>" × 34", and the minimum scratch pad size, according to company policy, was 3.5" × 5.75". What is the maximum number of scratch sheets that the supervisor could obtain from one stock sheet of the bond paper?   _____

15. A supply of 8½" × <u>11</u>" laser printer and copier paper needed to be cut for use in the front office of a university academic department. What percentage of waste would occur if this paper were cut from stock sheets that were 17½" × <u>22½</u>" in size?

_____

 ## Unit 34 DETERMINING PAPER NEEDS FOR IMAGED PRODUCTS

BASIC PRINCIPLES OF DETERMINING PAPER NEEDS FOR IMAGED PRODUCTS

It is extremely important for the graphic communications company estimator, press operator, and finishing personnel to correctly calculate the number of sheets of paper or other substrate needed for a given job. The information is needed because the estimator must know the amount of materials that will be used for a job so the correct estimate or bid can be given. In addition, the press operator must have available the correct number of sheets so he/she can print the job including the *makeready* and *spoilage*. Also, the finishing personnel need extra printed sheets to complete the job when they are required to perform different functions such as folding, trimming, and other finishing and binding functions.

Makeready involves using sheets to set up the equipment such as the printing press and paperfolder so the sheets will feed through the machines correctly. Spoilage means the sheets that are "consumed" during the printing on the press or the folding on the paperfolder when sheets do not feed correctly or are printed incorrectly for some reason, causing the press operator or paperfolder operator to destroy them. The skill and experience of the press operators and bindery personnel greatly determine the needed makeready and spoilage percentage. An amount frequently used throughout the industry for press and bindery makeready and spoilage is 10% for 1,000 or fewer finished copies and 5% for anything over 1,000 finished copies. For example, if a customer orders 8,000 brochures, 400 extra sheets (8,000 × .05 = 400) will be made available for the press and bindery personnel to use for setting up their equipment (makeready) and waste during the equipment run (spoilage). This means that a total of 8,400 sheets of paper should be made available prior to the press run so there will be 8,000 finished copies of the job for the customer.

Printed sheets are often *gang run*. This means that two or more copies of the printed product are included on one sheet when printed and then cut apart after the ink has dried (Figure 34–1).

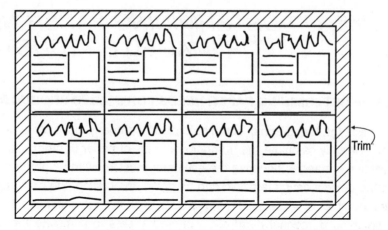

Figure 34–1   Gang running printed sheets on a single press sheet saves press time and handling time. This procedure also contributes to paper savings because fewer press sheets are needed for the spoilage factor.

Gang running saves considerable press time and handling time, thus greatly reducing production costs. Being able to calculate the number of printed sheets that can be included on one press sheet is an important responsibility of the estimating and production personnel. For example, if the printing press was large enough so that a 17½" × 22½" sheet could be printed at one time, it would save production and handling time to print four 8½" × 11" letterheads and then cut them apart after the ink has dried. The following examples help to demonstrate this principle.

Example 1:   Gang run—four finished sheets per press sheet:

Number of finished letterheads requested = 10,000

Gang run at four per sheet, 10,000 ÷ 4 = 2,500 press sheets × 5% for makeready and spoilage = 125 extra press sheets for a total of 2,625 sheets of paper 17½" × 22"

Productive press speed = 5,000 sheets per hour, thus 2,625 ÷ 5,000 = .525 of an hour or 31.5 minutes to run this job

Example 2:   Single sheet run—one finished sheet per press sheet:

Number of finished letterheads requested = 10,000

Sheet run at one per sheet, 10,000 ÷ 1 = 10,000 press sheets × 5% for makeready and spoilage = 500 extra press sheets for a total of 10,500 sheets of paper 8½" × 11"

Productive press speed = 5,000 sheets per hour, thus 10,500 ÷ 5,000 = 2.1 hours or 126 minutes to run this job

Often several pages of a book or magazine are printed on one large press sheet and then folded to form smaller pages. When this is done, each press sheet containing several pages is called a *signature.* Common signature page counts equal 4, 8, 12, 16, and 32. Sometimes 64 pages are included in a signature, but this is done only in special cases. Just as with gang running, this procedure saves considerable production and handling time.

It is very important to correctly calculate the amount of paper needed for a given job. The following two examples should help in understanding this procedure.

Example 1: How many stock sheets of 17" × 22" 20-pound bond paper are needed for a job requiring 1,200 finished 8½" × 11" letterheads?

1,200 × 5% = number of press sheets for makeready and spoilage

1,200 × .05 = 60 for a total of 1,260 press sheets

Number of pieces obtained from each stock sheet = 4

2 × 2 = 4 pieces from each sheet of paper 17" × 22"

$$\frac{17" \times 22"}{8\frac{1}{2}" \times 11"} \quad \text{meaning} \quad \frac{17" \div 8\frac{1}{2}" = 2}{22" \div 11" = 2} \quad \text{Thus } 2 \times 2 = 4 \text{ pieces}$$

1,260 ÷ 4 = 315 stock sheets are needed for the 1,200 finished letterheads

Example 2: How many stock sheets 25" × 38" are needed for 4,000 copies of a 96-page book that contains an untrimmed page size of 6" × 9"? The book will contain six, 16-page signatures that are 25" × 19" in size (Figure 34–2).

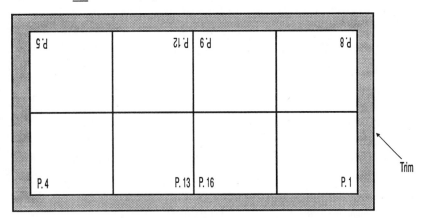

Figure 34–2 An example of a press sheet that contains a 16-page signature (8 pages per side) of a 6" × 9" untrimmed 96-page book.

4,000 × 5% = 200 sheets per side of each signature for makeready and spoilage

one signature × 2 sides = 400 sheets per signature for makeready and spoilage

4,000 + 400 = 4,400 sheets per signature

4,400 × 6 signatures = 26,400 press sheets

<u>38</u>" × 25" stock sheet ÷ 25" × <u>19</u>" press sheet = 2 (see Figure 34–2)

26,400 press sheets ÷ 2 press sheets per stock sheet = 13,200 stock sheets

Note: If the calculated number of stock sheets should equal a partial sheet, it is important to round up to a full number.

PRACTICAL PROBLEMS

1. A group of high school teachers wanted to show their students how flash cards could be used at an athletic event to present visual messages to fans in the stadium. They needed a total of 2,000 pieces of 110-pound index paper of various colors, 5" × <u>8</u>" in size. They contacted the local paper supply company and obtained the cut paper. Because there was no press or finishing equipment involved, there was no need to calculate makeready and spoilage. How many stock sheets of index paper, 25½" × 30½", were needed to fill this request?   _____

2. For a display that was being created in the school cafeteria, the high school teachers in problem 1 needed 750 pieces of 110-pound index paper of various colors, 5" × 7" in size. For this paper request, grain was not a factor, as the cards were going to be fastened to the existing display boards with staples. They went to the same paper supply company and acquired the cut paper. Again, there were no presses or finishing equipment involved, thus no makeready and spoilage needed to be considered. How many stock sheets of index paper, 25½" × 30½", were needed to fill this request?   _____

3. A customer ordered 5,000 copies of a one-color, 8½" × <u>11</u>" letterhead. He requested that the paper be 20-pound, 25% cotton content, white, and watermarked. For production purposes, the letterhead will be printed one-up on a small press, and the stock paper will be 17" ×<u>22</u>" in size. How many stock sheets will be needed to produce 5,000 finished sheets of letterhead using the standard makeready and spoilage guideline?   _____

4. The same customer who ordered the 5,000 one-color letterheads in problem 3 decided that he needed another letterhead. The quantity was the same (5,000), but he wanted this letterhead in two colors. Fortunately, the graphic arts company where the first one-color letterhead was printed had a two-color small press. Because the press operator was so skilled, the

makeready and spoilage factor was essentially the same as when he printed in one color. Using basic size bond paper, 17" × 22", how many stock sheets were needed to produce the 5,000 finished sheets of two-color letterhead?

    _____

5. For an order of 10,000 letterheads, 8½" × 11", one color, the graphic arts company production supervisor decided to have the operator of the 12" × 18" litho press be responsible for printing this job. With this press, it is possible to gang run two 8½" × 11" letterheads on one press sheet size. The selected paper was 20-pound, off-white, #1 grade sulfite in a stock sheet size of 17½" × 22½", thus the press sheets could be 17½" × 11¼" in size. The extra ½" and ¼" of paper on the two sides permits some trimming after the letterhead is printed. This gives a neat and clean appearance to the edges of the finished letterhead. How many stock sheets, using the standard makeready and spoilage amount, will be needed to produce 10,000 finished letterheads using this production procedure?

    _____

6. The manager of the in-plant graphic communications facility established a standard procedure for printing one-color and multiple-color, 3½" × 2" business cards (Figure 34–3). This layout of ten cards per 7½" × 11" press sheet provided sufficient trim on all four sides so the cards would be neat and clean. She also found that the basic makeready and spoilage factor worked well for her press operators. How many 7½" × 11" press sheets are needed for 500 business cards?

    _____

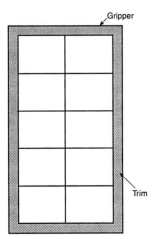

Figure 34–3   The layout for gang running ten 3½" × 2" business cards on a 7½" × 11" press sheet.

7.   For another job, the in-plant manager from problem 6 was asked to produce 1,250 business cards. How many standard press sheets are needed for this quantity of business cards?   _____

8.   Some single-color invoice forms measuring 5½" × 8½" were being gang run, 4-up, on a press sheet measuring 11" × 17". The order called for 15,000 invoice forms, and the customer specified that the paper should be 16-pound canary bond. The production manager selected the 17" × 22" paper from his stockroom and prepared the correct number of press sheets for the press operator based on the standard makeready and spoilage factor. How many stock sheets were needed for this job?   _____

Note: The following specifications will serve as the basis for problems 9–13.

At a packaging company, the supervisor of the design and layout department created a package layout for a can of mixed vegetables. When placed flat, each individual can wrap measured 4¼" × 10½". The 4-color packaging was going to be printed on a 5-color, sheet-fed litho press so that a clear varnish coating could be placed over the process color printed sheets. On this press, the skilled operators were able to keep the makeready and spoilage at the rate of 6% for any job with a minimum of 5,000 press sheets. The selected paper was a coated book, basic size of 25" × 38", and the press sheet size would be 25" × 19", thus the can package wrap could be gang run.

9.   How many finished sized packaging sheets can be gang run on one press sheet?   _____

10.   Based on producing 175,000 package can wraps, how many press sheets will be needed not considering makeready and spoilage?   _____

11.   Based on producing 175,000 package can wraps, how many press sheets will be needed when makeready and spoilage are considered?   _____

12.   What number of stock sheets are needed to complete this job?   _____

13.   With the productive press speed of 5,000 sheets per hour, how many hours and minutes should be scheduled to print this job?   _____

Note: The following specifications will serve as the basis for problems 14–17.

The specifications for a 16-page booklet were listed as follows: 6" × 9" pretrimmed size, 2,500 copies, one color, uncoated book paper 25" × 38". The graphic arts production personnel decided to run the job on their 18" × 12" litho press, thus it would be possible to produce this job with two 8-page signatures. The standard makeready and spoilage percentage worked well in this plant.

14. How many total press sheets were needed for this job not considering makeready and spoilage?   _____

15. How many total press sheets were needed for this job counting the standard makeready and spoilage?   _____

16. How many stock sheets were needed to produce the 2,500 copies of the 16-page booklet?   _____

17. With the productive press speed of 5,000 sheets per hour, how many hours and minutes should be scheduled to print this job?   _____

Ink Needs

SECTION

9

Ink is the silent messenger. It is the substance that makes it possible for us to read the words, see the illustrations, and enjoy the photographs. Without ink, the pages of books, magazines, and newspapers would be blank. Some people might state that they would enjoy looking at a "blank" textbook, such as this textbook, but then there would be no reason to print and bind the book.

Printing ink is a *commodity* that constitutes a sizeable portion of the manufacturing industry. There are approximately 300 printing ink manufacturing companies in the United States and many more throughout the industrial nations of the world. There are thousands of people employed in the printing ink industry.

Because printing ink is so important in the graphic communications industry, people need to know how to calculate the amount of printed area that a specific amount of ink will cover. It takes a considerable amount of research and dedicated study to accurately determine how many square inches of area can be covered with a single pound of printing ink. The graphic communications estimator needs to know how to mathematically utilize this data when making cost estimates and final bid specifications. In addition, the estimator, and anyone else involved in determining how much ink will be needed for a given printed product, needs to know the mathematical methods for calculating the amount of ink for specific jobs. It certainly would be unproductive to run out of ink when the press operator is 90% finished with a long run of several thousand copies. This would be a costly error that could cause a company to have financial problems if the error was duplicated too many times. The procedures for determining ink coverage, ink needs, and ink costs are not difficult, but the calculations must be completed with accuracy or they are of little value. The basic principles information and the several problems in each unit have been prepared to make you aware of this important area and to give you some knowledge and skills that just might provide you with a possible area of employment.

Unit 35 CALCULATING INK COVERAGE

BASIC PRINCIPLES OF CALCULATING INK COVERAGE

Measuring the amount of ink needed to print images on paper or any of the many substrates is somewhat difficult and inaccurate even when the best known methods are used. It is, though, important to make every attempt to calculate ink coverage so that the correct amount of ink can be made available for a given job and so that costs can be assigned to the ink needed for a given job.

It is relatively easy to determine how much paper or other substrates will be needed for a given job because the sheets of paper or number of items to be printed can be counted. On the other hand, there are several variables when determining the amount of printing ink that will be needed for a given job. These major variables include the following: (1) the size of the form or overall image area, determined in square inches; (2) the percentage of the image area being printed on the paper or substrate; and (3) the number of finished copies, including equipment makeready and spoilage. The formula used to calculate ink coverage includes these three stated variables (Figure 35–1). This formula serves a definite

Ink Coverage Forumula

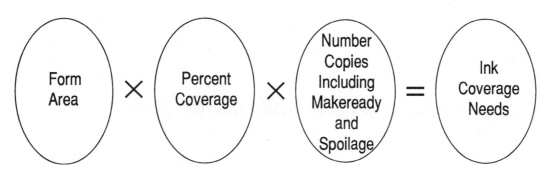

35–1 The formula for determining the ink coverage needs for a specified printed product.

value and helps the estimator closely calculate the correct amount of ink needed for specified jobs. There is, though, another "tool" that makes it possible to establish some consistency in estimating the amount of ink needed for a given job. This is the *ink coverage chart* (Figure 35–2). In using this chart, the estimator has to make a decision about the image area that will be covered with ink on a given job. This is a critical variable because selecting a specific percent of coverage over another does make a difference in the amount of ink that is estimated for a given job.

Lithographic Ink Coverage Chart

Type of form	Percent of coverage
Very light composition, no halftones	15
Normal composition, no halftones	20
Normal composition, bold paragraph heads	25
Medium composition, no halftones	35
Heavy composition, no halftones	50
Halftones	50
Screen tints*	—
Solids	100
Reverses	60–90
Process color separations:	
Magenta and cyan	40
Yellow	50
Black	20

*Approximate screen percentage

Figure 35–2   The general guidelines for determining lithographic ink coverage based on the content of the page. (Credit: Ruggles, *Printing Estimating*, 4th ed.)

The following two examples are representative of how the amount of ink coverage can be predetermined for a job. This procedure can be used with any of the imaging methods as long as the "ink coverage chart" is appropriate for the process.

Example 1:   How many square inches of ink coverage is required for a job with the following specifications?

Number of pages = 1; form (image) area per page = 5" × 8½"; ink coverage per page = 20%; number of copies = 50,000; equipment makeready and spoilage = 5%

5" × 8½" (image size) = 42.5 square inches per page

42.5 × .20 (coverage) = 8.5 square inches per page

50,000 × .05 (makeready and spoilage) = 2,500 press sheets

50,000 copies + 2,500 copies for makeready and spoilage = 52,500 press sheets

8.5 × 52,500 = 446,250 square inches of ink coverage

Example 2: How many square inches of ink coverage will be required for a job with the following specifications?

Page size = 8½" x 11"; form (image) area per page = 40 x 54 picas; number of pages = 2, front-and-back (duplexed) on one sheet; ink coverage page one = normal composition, bold paragraph heads; ink coverage page two = heavy composition, no halftones; number of finished copies = 35,000; normal makeready and spoilage = 5%

40 x 54 picas (image size) = 2,160 square picas

2,160 ÷ 36 (square picas per square inch) = 60 square inches per page

60 x .25 (ink coverage of page one) = 15 square inches page one

60 x .50 (ink coverage of page two) = 30 square inches page two

35,000 x .05 (makeready and spoilage per side) = 1,750 sheets per run

35,000 + 1,750 = 36,750 sheets per page (each side)

15 x 36,750 = 551,250 total square inches of ink coverage for page one

30 x 36,750 = 1,102,500 total square inches of ink coverage for page two

551,250 + 1,102,500 = 1,653,750 square inches of ink coverage for pages one and two

PRACTICAL PROBLEMS

1. John, a litho press operator, checked the ink supply and noted that there was no opaque red available. He knew of a job that had just been sold that required opaque red ink, thus he had to determine the square inches of coverage so the purchasing supervisor could place an order. He obtained the job specifications and found the following information: number of pages = 1; form area = 5" x 8½"; very light composition, no halftones; 75,000 copies plus or minus 5,000 copies, thus no makeready and spoilage had to be considered. How many square inches of ink coverage is needed for this job? _____

2. A small calendar is to be printed, but because of the small size, it is possible for the copy preparation person to place ten calendars on one page for a gang run. The estimator had suggested this procedure, thus it is possible to calculate the needed ink coverage for the job. This calendar job calls for the following specifications: one page; one sheet; form area for all ten calendars = 5.5" x 9.25"; normal composition, no halftones; 100,000 finished calendars; and 3% makeready and spoilage. How many square inches of ink coverage will be required for this calendar job? _____

3. A political campaign leaflet had been sold to a candidate for a state office. This was a sizeable job for the graphic communications company. The competition for this bid had been strong, thus the estimator was very careful to make all of her calculations as accurately as possible. The estimator was given these specifications: one page, one sheet; page size = 8½" × 11"; image area = 7½" × 9½"; heavy composition with one halftone; 250,000 copies; and 2% makeready and spoilage. How many square inches of ink coverage will be required for this political leaflet?   _____

Note: The following specifications are to be used for problems 4–7.

The president of a large package printing company obtained the contract for a new cereal box. The box was small, but the quantity was substantial. If the production quality was maintained, this could be a long-standing contract which the president wanted to retain. The specifications for the first order were as follows: 500,000 finished boxes; flat image area of the box before folding and gluing = 19" × 25"; 100% process color; 3% makeready and spoilage for the 4-color press and box-making equipment.

4. How many square inches of coverage will be required for the Cyan ink?   _____

5. How many square inches of coverage will be required for the Magenta ink?   _____

6. How many square inches of coverage will be required for the Yellow ink?   _____

7. How many square inches of coverage will be required for the Black ink?   _____

8. The high school biweekly newspaper was in the process of being re-bid. Several regional newspaper publisher-printers bid on the job, as this was a profitable contract. The specifications were listed as follows: 8 pages per issue; 16 issues per academic year; tabloid size; image area per page = 10.25" × 12.75"; 2,000 copies per issue; and the average ink coverage per page = 40%. The estimator for the low bid used a 2% makeready and spoilage figure for each issue. How many square inches of ink will be required for the 16 issues of the high school newspaper based on these specifications?   _____

9. There were several bidders for the high school biweekly newspaper detailed in problem 8. The estimator for the next-to-the-lowest bidder prepared all the calculations correctly but used a makeready and spoilage figure of 3.5% for each issue. How many square inches of image coverage will be required for the 16 issues of the high school newspaper based on this one specification change? _____

Note: The following specifications are to be used for problems 10–12.

An instruction manual was being planned by the Director of Training of a large manufacturing company. The training director, in cooperation with the Director of Publications, prepared the specifications for the job. The specifications included the following information: 25 pages, printed one side; 8½" × 11" page size with margins of ¾ inch on each of the four sides; medium composition, no halftones; and one color. Estimates were requested on quantities of 2,000; 3,000; and 5,000.

10. What is the estimated square inches of image coverage for 2,000 copies using a makeready and spoilage factor of 5%? _____

11. What is the estimated square inches of image coverage for 3,000 copies using a makeready and spoilage factor of 2.5%? _____

12. What is the estimated square inches of image coverage for 5,000 copies using a makeready and spoilage factor of 2%? _____

13. The owners of a Chinese restaurant were running low on table place mats. They decided to have a quantity printed so they would be ready for the dinners and various celebrations surrounding the annual Chinese New Year parties. Their place mat was of two colors, transparent red and transparent yellow. It was 13½" × 9¾" in size; the red image bled off the four edges, and the yellow image size was 12¾" × 9". The red image covered approximately 45% of the place mat, and the yellow image covered approximately 10% of the place mat. They placed an order for 25,000 place mats. They did not request an estimate of the cost, as they had utilized the services of the Cedar Valley Graphic Communications Company before and fully trusted their pricing. How many square inches of red image area were required for this job assuming that a 5% makeready and spoilage factor was selected because a two-color press was going to be used to produce the place mats? _____

14. Besides the red ink for the place mat job described in problem 13, the yellow ink image area also had to be determined. How many square inches of yellow image area were there to cover for this job based on the specifications stated in problem 13?   _____

15. The owner of a screen printing company accepted a job for 5,000 computer mouse pads that contained illustrations and type matter in three colors. The image area was 8" × 6½". The estimator looked at the comprehensive layout and estimated that the red image covered 10% of the image area, the green image covered 5%, and the black image covered 15%. A multicolor, automatic screen printing press was going to be used, thus a 3% makeready and spoilage factor was selected. How many total square inches of ink coverage were needed for this job?   _____

Unit 36   DETERMINING INK NEEDS AND COSTS

BASIC PRINCIPLES OF DETERMINING INK NEEDS AND COSTS

Usually, ink costs are a small portion of most printed jobs, but the cost of ink is substantial enough to make certain that it not be overlooked when estimating the total cost of a printed product. The basic procedures for determining the square inches of image area for one or more ink colors were fully covered in the previous unit. In this unit, there will be sufficient opportunity to gain experiences in calculating the amount of ink needed based on the image area to be covered (Figure 36–1).

Ink Need Forumula

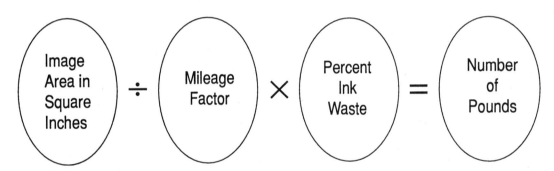

Figure 36–1   The formula for determining the number of pounds of ink after the total square inches of image area have been calculated.

Another critical variable is the *ink mileage factor* of the ink (Figure 36–2). Different types and colors of ink will cover differing amounts of substrate area according to the type of substrate being printed upon. The different ingredients of the various inks make a major difference in their mileage factors. With the many kinds and types of papers, there is also a major difference in how much area ink will cover from one paper to another. Mileage factor charts that have been prepared by ink manufacturing company research personnel are valuable when determining the amount of ink needed for a given job.

Lithographic Ink Mileage Schedule Chart

Type of Stock		Color of Ink							
	Regular Black	Rubber Base Black	Purple	Trans-parent Blue	Trans-parent Green	Trans-parent Yellow	Chrome Yellow	Persian Orange	Trans-parent Red
No. 1 enamel	425	445	360	355	360	355	285	345	350
No. 2 enamel	400	435	355	345	355	355	270	335	345
Litho coated	380	425	350	340	350	355	260	325	345
Dull coated	375	415	320	335	335	340	250	310	340
M.F. book	400	435	350	340	350	340	250	325	340
Newsprint	290	350	250	230	250	235	165	240	240
Antique	275	335	235	220	235	220	150	225	225

Type of Stock		Color of Ink				Process Colors		
	Opaque Red	Brown	Over-print Varnish	Opaque White	Tint Base	Cyan	Magenta	Yellow
No. 1 enamel	350	345	450	200	400	355	350	355
No. 2 enamel	345	340	435	185	390	350	347	355
Litho coated	340	335	425	175	380	340	345	355
Dull coated	325	325	415	165	375	335	340	340
M.F. book	340	335	425	175	385	340	340	340
Newsprint	190	240	—	150	265	235	240	235
Antique	175	225	—	135	250	220	225	220

Type of Stock	Color of Ink		
	Silver	Met-tallics	Fluor-escent
No. 1 enamel	335	215	270
No. 2 enamel	—	—	—
Litho coated	300	200	240
Dull coated	285	200	240
M.F. book	—	—	—
Newsprint	—	—	—
Antique	220	130	160

Note: Numbers given indicate thousands of square inches of coverage per pound of ink

Figure 36–2 A lithographic ink mileage schedule for determining coverage of several ink colors and types of paper.

Ink waste, both on the press and in the container, is difficult to measure either in advance or while it is happening. Over the years, graphic arts estimators and production personnel have determined that there is a range of 2% to 5% ink waste that can be assumed with each job. A constant ink waste percentage figure should be established in all graphic arts facilities where printing presses are used. It is important to keep this percentage as low as possible. For the problems in this unit, a constant waste factor of 3% will be used when calculating the amount of ink needed for a given job.

To review, there are three important numbers that are needed before it is possible to calculate the pounds of ink that will be required for a job (see Figure 36–1). These are the image area in square inches (see Unit 33), the mileage factor of the ink according to the type of paper being used (see Figure 36–2), and the ink waste factor. An example of the procedure for determining the number of pounds of ink needed for a given job is shown as follows.

Example: Specifications of the job: square inches of image area = 425,000; ink mileage factor = 380,000 sq in (regular black ink on litho coated paper); ink waste factor = 3%

425,000 ÷ 380,000 × .03 = number of pounds

425,000 ÷ 380,000 = 1.11842 (carry to five places and round up if needed)

1.11842 × 1.03 = 1.15197 pounds of ink

There also will be opportunities to calculate the cost of ink for given jobs in this unit. The basic formula involves three pieces of information: (1) the number of pounds of ink, (2) the selling price of the ink, and (3) the shipping cost of the ink (Figure 36–3). Ink costs vary according to the ink type, ink color, and amount of ink purchased at one time (Figure 36–4).

Ink Cost Forumula

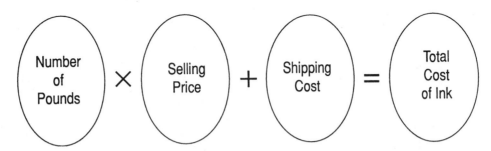

Figure 36–3   The formula for determining the total cost of ink after the number of needed pounds has been calculated.

A Typical Pricing Schedule
of
Lithographic Inks for Small Presses

Color	Rubber Base	Double Rubber Base	Omni	Silverline	1 Lb. Cans	5 Lb. Cans (per Lb.)
Black	29995	22X	80800	96000	8.60	7.50
Pantone Yellow	RY2200	5650	80801	96001	13.20	12.00
Pantone Warm Red	RR2400	6242	80802	96002	14.90	14.00
Pantone Rubine Red	RR2525	7678	80803	96003	14.90	14.00
Pantone Rhodamine Red	RR2575	7679	80804	96004	16.55	15.65
Pantone Purple	RP2100	7680	80805	96005	16.55	15.65
Pantone Reflex Blue	RB2850	3834	80806	96006	15.20	14.00
Pantone Process Blue	RB2650	4342	80807	96007	14.70	13.75
Pantone Green	RG2300	4921	80808	96008	16.35	15.30
Pantone Rans. White	RB1000	0023	80810	96009	9.10	8.10
Pantone Mixing Black	RB2950	9958	80809	96010	8.60	7.50
Pantone Violet	013RB	3855	0130M	96013	16.55	15.65
Washington Yellow	RB11	XX11	OM11	ST11	13.65	12.60
Adams Orange	RB12	XX12	OM12	ST12	14.60	13.55
Jefferson Orange	RB13	XX13	OM13	ST13	15.00	14.00
Madison Orange	RB14	XX14	OM14	ST14	15.35	14.35
Monroe Red	RB15	XX15	OM15	ST15	14.60	13.55
Jackson Red	RB16	XX16	OM16	ST16	14.60	13.55
Van Buren Red	RB17	XX17	OM17	ST17	14.60	13.55
Harrison Wine	RB18	XX18	OM18	ST18	15.55	14.35
Tyler Purple	RB19	XX19	OM19	ST19	15.65	14.60
Polk Blue	RB20	XX20	OM20	ST20	15.25	14.20
Taylor Blue	RB21	XX21	OM21	ST21	14.90	13.90
Fillmore Blue	RB22	XX22	OM22	ST22	14.75	13.60
Pierce Blue	RB23	XX23	OM23	ST23	15.30	14.25
Buchanan Aqua	RB24	XX24	OM24	ST24	15.50	14.50
Lincoln Green	RB25	XX25	OM25	ST25	14.90	14.00
Johnson Green	RB26	XX26	OM26	ST26	16.05	15.00
Grant Green	RB27	XX27	OM27	ST27	15.20	14.40
Hayes Olive	RB28	XX28	OM28	ST28	14.70	13.65
Garfield Green	RB29	XX29	OM29	ST29	15.35	14.20
Arthur Ochre	RB30	XX30	OM30	ST30	14.35	13.20
Cleveland Mustard	RB31	XX31	OM31	ST31	14.35	13.20
Hoover Orange	RB32	XX32	OM32	ST32	14.35	13.20
McKinley Tan	RB33	XX33	OM33	ST33	14.40	13.25
Roosevelt Brown	RB34	XX34	OM34	ST34	14.50	13.30
Taft Brown	RB35	XX35	OM35	ST35	14.60	13.55
Wilson Brown	RB36	XX36	OM36	ST36	14.75	13.60
Harding Brown	RB37	XX37	OM37	ST37	14.65	13.50
Coolidge Grey	RB38	XX38	OM38	ST38	13.75	12.65
Opaque White	10200	0024	OM39	ST39	9.10	8.10
All Designer Colors	RB40–49	XX40–49	OM40–49	ST40–49	16.50	15.50

Figure 36–4   An example of a lithographic ink pricing schedule.

The basic procedure for determining the cost of ink for a specified job is shown in the following example.

Example: Specifications of the job: number of pounds of ink needed = 1.15197; color of ink = Garfield Green; selling price per pound, purchased in one pound amounts = $15.35; shipping cost = $5.50

1.15197 × $15.35 + $5.50 = total ink cost

1.15197 × $15.35 = $17.68 (round at two places)

$17.68 + $5.50 = $23.18 total cost of ink for this job

PRACTICAL PROBLEMS

1. A graphic designer selected a purple ink and an antique finish paper for a job on which she was working. How many square inches of image will purple ink cover when it is used with antique finish paper? _____

2. Another graphic designer was working on a brochure for a university where the school colors were purple and gold. He looked at the ink mileage schedule but could not find an ink specified as gold in color. He made a decision to use the specifications for chrome yellow as being representative of a gold-colored ink. How many square inches of image will chrome yellow ink cover if a number 1 enamel paper is used? _____

3. The graphic designer described in problem 2 was exploring the possibilities of using either a number 1 enamel paper or a machine finish (M.F.) book paper. He decided to submit his recommendation that either type of paper would serve the purpose as intended. If the budget was critical, the estimator could bid both papers and that just might make the difference in their company gaining the contract. What is the numerical difference between the two papers in the mileage factor when chrome yellow ink is used? _____

4. Ink pricing and cost schedules are important for the ink manufacturer, and they are important for the estimator of a graphic communications company. Using the cost schedule in Figure 36–4, how much does one pound of Roosevelt Brown ink cost? _____

5. Often there is need to purchase more than one pound of ink at a time. How much will three pounds of Washington Yellow ink cost (total) when it is purchased in one pound cans?

    _____

6. Gregg, a press operator for a "jiffy walk-in" graphic communications outlet, needed some Tyler Purple litho ink for a job that had been sold. It was a sizeable job, thus he decided to purchase five pounds of the ink at one time. What is the total cost of the five pounds of Tyler Purple ink?

    _____

7. A wise purchasing agent will "gang" purchase ink whenever possible to save money. Financial managers of some ink companies will permit "five-pound" pricing even when two different colors of ink are purchased at the same time. An example would be that three pounds of Pantone Rubine Red and four pounds of Pantone Green were ordered at the same time. The usual policy is that the price per pound is calculated based on the higher cost of the two inks according to the five-pound, per pound price. What would be the cost of three pounds of Pantone Rubine Red and four pounds of Pantone Green based on the five-pound pricing policy?

    _____

8. There was a small leaflet to be printed in one color, but the quantity was substantial. The estimator made the calculations and determined that there would be 873,452 square inches of image area to be covered in this job. The selected ink was a standard brown, and the paper was classified as dull coated book. How many pounds of ink will be needed for this job?

    _____

9. A weekly newspaper business manager was concerned about the cost of ink. She did some research and found that regular black ink was being used to print their eight-page weekly newspaper on an average grade of newsprint paper. She suggested that the press operator change from regular black ink to rubber base ink. How many more square inches of image area will rubber base ink cover than regular black ink?

    _____

10. For the eight-page weekly newspaper described in problem 9, 30,000 copies of each issue were printed. On average, each page contained 131 square inches of image. Using this information, each issue included 31,440,000 square inches of image area. Using the mileage factor for regular black ink and the 3% ink waste factor, how many pounds of ink are used for each issue of the weekly newspaper?

    _____

11. As pointed out in problem 9, the business manager was interested in looking into using rubber base ink instead of regular black ink. In addition, it was shown in problem 10 that each issue of the weekly newspaper included 31,440,000 square inches of image area. Using the mileage factor for rubber base black ink and the 3% ink waste factor, how many pounds of ink would be used for each issue of the weekly newspaper if the change were made?

_____

12. A magazine publisher looked into the cost of the ink for each edition of his weekly magazine. He found that 35.66 pounds of regular black ink, on average, was being used for each issue. In addition, he found that the average shipping cost per pound was $1.45. What was the total cost of the ink, including shipping, for each issue of the magazine?

_____

13. There were three litho presses in an average-sized graphic arts company in a city of 35,000 people. The production manager noticed that each week press operator A used an average of five pounds of Polk Blue ink, press operator B used an average of three pounds of Lincoln Green ink, and press operator C used an average of four pounds of Harding Brown ink. He was told by the business manager that the per pound shipping cost each week averaged $2.05. What was the average total cost of ink, including the shipping, per week for the three presses?

_____

14. A job required 3.67 pounds of Van Buren Red and 4.26 pounds of Fillmore Blue. The ink was usually purchased in large amounts and kept on the shelf for ready use, but the estimator always prepared estimates based on per pound prices. Also, the estimator calculated a constant shipping cost of $1.75 per pound. What was the total cost of the ink for this job including shipping?

_____

15. Aaron, a health care management official, ordered 40,000 flyers printed on litho coated paper that included two colors of ink—Pantone Reflex Blue and Jackson Red. Both ink colors can be considered transparent colors. The total square inches of blue image area for the 40,000 flyers including makeready and spoilage was 759,780. The total square inches of red image area for the flyers was 943,639. Barton, the graphic arts company estimator, calculated the total cost of the two colors of ink. He based his figures on the information found in the ink mileage schedule chart, the pricing chart, a 3% ink waste, and an average per pound shipping cost of $1.25. The graphic communications company executive assistant, Seth Byron, looked at the ink cost figures and exclaimed surprise at what he saw. What total ink cost figure was presented to Seth?   _____

Costing Materials and Services

How much will it cost to cut and handle the paper for 5,000 finished sheets of 8½" × 11" letterhead? In addition, how much will it cost for the 20-pound bond paper needed to produce the 5,000 finished sheets of letterhead? These questions and others must be answered by estimators and management personnel so they can perform the needed calculations to provide the requested pricing of finished products that are manufactured in graphic communications companies.

Labor costs for any finished product are usually more than the actual cost of the material or finished product. For example, the cost of producing 10,000 sixteen-page booklets always includes a high percentage of labor costs. This begins at the design and layout stage and continues through the copy preparation, photo/electronic image conversion, image carriers, image transfer, and finishing and binding stages. Many people are needed to provide all of the services performed along the production route. The information and problems provided in Unit 35 are directed toward the aspect of labor costs for performing a required service. This information and opportunities for making some basic mathematical calculations are only examples of the entire costing procedures that are utilized throughout the graphic communications industry. The ability to look at and correctly use information in pricing charts also makes a person valuable to the management in a company.

Understanding and making correct calculations when it comes to determining the cost of hundreds and thousands of sheets of paper are important qualities of an estimator. An estimator must know the graphic communications production processes as well as know and use the basic mathematical procedures. Using paper catalogs and pricing sheets are everyday tasks of estimators and management personnel of most manufacturing companies. The information and problems in Unit 36 should help you understand the basics of making the correct calculations when it comes to determining the cost of paper. Look closely at the calculation procedures and learn how to use the information provided in paper catalogs. This knowledge may help you find some challenging and exciting employment for a life-long career in the graphic communications industry.

Unit 37   DETERMINING COST OF CUTTING AND HANDLING PAPER STOCK

BASIC PRINCIPLES OF DETERMINING COST OF CUTTING AND HANDLING PAPER STOCK

An important aspect of estimating costs is the charge for cutting the paper. Paper is cut by the *lift*. A lift is the maximum number of sheets of paper stock that can be properly handled at one time either by hand or by paper-handling devices. Sometimes a lift is defined as the amount of paper that can be placed under the clamp of a paper cutter at one time. This is not always a good definition because this amount of paper may simply be too much to properly handle by either hand or machine.

Paper Cutting Cost Schedule

Table One

Bond, Ledger, Book, Newsprint

500 sheets to a lift.
If only 250 sheets can be cut to a lift, add 100%.
If 1,000 sheets can be cut to a lift, deduct 50%.

Sq. In. to Finished Pc.	Number of Finished Size Pieces											Add'l 1 m
	100	500	1m	2m	3m	5m	10m	15m	25m	50m	100m	
20	$1.40	$1.70	$2.10	$2.75	$3.40	$4.70	$7.90	$11.15	$17.55	$33.65	$65.80	$.64
30	1.40	1.75	2.15	2.85	3.55	4.90	8.25	11.65	18.40	35.35	69.20	.68
40	1.40	1.75	2.15	2.95	3.65	5.05	8.60	12.15	19.25	37.05	72.55	.71
50	1.45	1.80	2.20	3.00	3.75	5.25	8.95	12.70	20.10	38.70	75.90	.74
60	1.45	1.80	2.25	3.10	3.85	5.40	9.30	13.20	20.95	40.40	79.30	.78
80	1.45	1.85	2.35	3.25	4.10	5.75	10.00	14.20	22.65	43.80	86.00	.84
100	1.45	1.90	2.40	3.40	4.30	6.10	10.70	15.25	24.35	47.15	92.75	.91
125	1.45	1.95	2.50	3.60	4.60	6.55	11.55	16.55	26.50	51.40	101.20	1.00
173	1.50	2.05	2.75	3.95	5.15	7.45	13.25	19.10	30.75	59.85	118.05	1.16
200	1.50	2.10	2.85	4.15	5.40	7.90	14.15	20.35	32.85	64.05	126.45	1.25
250	1.55	2.20	3.05	4.55	5.95	8.80	15.85	22.95	37.10	72.50	143.30	1.42
350	1.60	2.45	3.45	5.30	7.05	10.55	19.30	28.05	45.60	89.40	177.00	1.75
400	1.60	2.55	3.65	5.70	7.60	11.45	21.05	30.65	49.85	97.85	193.85	1.92
500	1.65	2.75	4.05	6.45	8.70	13.20	24.50	35.75	58.35	114.75	227.55	2.26
700	1.75	3.20	4.90	7.95	10.90	16.75	31.40	46.05	75.30	148.50	294.90	2.93
800	1.80	3.40	5.30	8.75	12.00	18.55	34.85	51.15	83.80	165.40	328.60	3.26
1000	1.90	3.85	6.10	10.25	14.20	22.05	41.75	61.45	100.80	199.20	396.00	394
1600	2.20	5.15	8.55	14.85	20.80	32.70	62.45	92.25	151.75	300.55	598.15	5.96
2000	2.40	6.00	10.20	17.90	25.20	39.80	76.30	112.75	185.70	368.10	732.90	7.30

Figure 37–1   A cutting charge table for bond, ledger, book, and newsprint papers.

200

Paper Cutting Cost Schedule

Table Two

Cardboard, Cover, Index Bristol, Tagboard

200 sheets to a lift.
If only 100 sheets can be cut to a lift, add 100%.
If 400 sheets can be cut to a lift, deduct 50%.

Sq. In. to Finished Pc.	Number of Finished Size Pieces											Add'l
	100	500	1m	2m	3m	5m	10m	15m	25m	50m	100m	1 m
15	$1.50	$2.15	$2.85	$4.20	$5.50	$8.05	$14.45	$20.85	$33.60	$65.55	$129.45	$1.27
20	1.50	2.15	2.90	4.30	5.65	8.30	14.90	21.50	34.70	67.70	133.70	1.31
30	1.55	2.20	3.05	4.50	5.90	8.70	15.75	22.75	36.80	71.90	142.10	1.40
40	1.55	2.25	3.15	4.70	6.20	9.15	16.60	24.05	38.90	76.10	150.50	1.48
50	1.55	2.30	3.25	4.90	6.45	9.60	17.45	25.35	41.05	80.35	158.95	1.57
60	1.55	2.40	3.35	5.10	6.75	10.05	18.35	26.60	43.15	84.55	167.35	1.65
80	1.60	2.50	3.55	5.45	7.30	10.95	20.05	29.15	47.40	93.00	184.20	1.82
100	1.60	2.60	3.75	5.85	7.85	11.80	21.80	31.75	51.65	101.45	201.05	1.99
125	1.65	2.75	4.00	6.30	8.50	12.95	23.95	34.95	56.95	112.00	222.10	2.20
150	1.70	2.85	4.25	6.80	9.20	14.05	26.10	38.15	62.30	122.60	243.20	2.41
175	1.70	3.00	4.50	7.25	9.90	15.15	28.25	41.35	67.60	133.15	264.25	2.62
225	1.80	3.25	5.05	8.25	11.25	17.35	32.55	47.80	78.20	154.25	306.35	3.04
300	1.85	3.70	5.80	9.65	13.35	20.70	39.05	57.40	94.10	185.90	369.50	3.67
350	1.95	3.95	6.30	10.60	14.70	22.90	43.35	63.80	104.75	207.05	411.65	4.09
450	2.05	4.50	7.35	12.55	17.45	27.35	52.00	76.65	125.95	249.25	495.85	4.93
600	2.25	5.30	8.85	15.40	21.60	34.00	64.95	95.90	157.80	312.60	622.20	6.19
700	2.40	5.85	9.90	17.30	24.35	38.40	73.55	108.75	179.05	354.85	706.45	7.03
900	2.65	6.95	11.95	21.15	29.85	47.25	90.85	134.80	221.50	439.30	874.90	8.71
1000	2.75	7.45	12.95	23.05	32.60	51.70	99.45	147.20	242.75	481.55	959.15	9.55

Figure 37–2   A cutting charge table for cardboard, cover, index-bristol, and tagboard papers.

The owner, manager, or even the CEO will establish the charges associated with cutting paper. There are standard cutting fees that have been developed on a national basis, but management personnel in each graphic arts plant must determine their own fees based on their own equipment and working conditions (Figures 37–1 and 37–2). Fee charts are excellent tools for the estimator when the costs of producing jobs are calculated. Without prepared charts that can either be referred to when making estimates or placed into computer estimating software programs, there will be little opportunity for consistency in estimating and pricing. In addition to the charges provided in the pricing table, there are rules for charging for special situations. Two such situations include the following: (1) for *cut apart only*, there should be a 40% discount from the table cost; and (2) for *back trimming*, it is appropriate to add 50% to the charge listed in the table. The term *cut apart only* means to obtain a given sheet size when it is only necessary to make one cut in the basic stock paper The term *back trimming* is used when all four sides of the newly cut sheets are trimmed as compared to using the precut sides of the stock sheets from the paper mill.

The cost of handling paper stock and other substrate before and after imaging is an expense that is usually charged against each finished job. Most graphic arts management personnel add 10% to the cost of the material used to cover the cost and expense of transporting, uncrating, unwrapping, shelf storage, seasoning, jogging, and conveying the material to the presses or imaging equipment. Sometimes the cost of cutting the substrate is included within the handling charge, but that is determined by the company management.

The method for determining the cost of cutting paper stock is straightforward; there are no secrets. There are three basic steps to determining the cost of cutting sheet paper stock. They are listed as follows:

1. Calculate the area in square inches of the finished paper size.

2. Find the number of pieces of paper stock that are to be cut on the cutting cost table (see Figure 37–1).

3. Read the number listed; that is the cost.

Several examples are given to provide reinforcement and understanding of these three steps. The basic rules of interpreting and using the tables are listed where appropriate in the following examples.

Example 1:   What is the charge for cutting 5,000 4" × 5" pieces of book paper from stock/parent sheets?

Area of each finished piece: 4" × 5" = 20 square inches

Cutting charge for 5M = $4.70

Example 2:   What is the cutting charge for 10,000 6¼" × 8¾" pieces of bond paper from stock/parent sheets?

Area of finished piece 6¼" × 8¾" = 54$\frac{11}{16}$ square inches. (**Note:** If the number of square inches in the finished piece is not on the table, it is appropriate to use the next highest figure.)

Cutting charge for 10M = $9.30

Example 3:   What is the cutting charge for 5,500 pieces of bond paper that is 5" × 8" in size?

Area of finished piece: 5" × 8" = 40 square inches

Cutting charge for 5M = $5.05. (**Note:** If the number of finished pieces is not on the chart, the cutting charge should be prorated.)

Cutting charge for 10M = $8.60

Cutting charge per 1,000 pieces over 5,000 pieces = ($8.60 minus $5.05) $\frac{1}{5}$ = .71

Cutting charge for 5,500 = $5.05 + .36 = $5.41

Example 4:   What is the charge for cutting 20,000 pieces that are 9" × 12" from basic size book paper?

Area of each finished piece: 9" × 12" = 108 square inches

Cutting charge for 15M = $16.55

Cutting charge for 25M = $26.50

Cutting charge per 1,000 over 15M = ($26.50 minus 16.55) $\frac{1}{10}$ = .995

Cutting charge for 20M = $16.55 + (5 × .995) = $21.53

PRACTICAL PROBLEMS

Note: For the conditions given in problems 1–15, the cutting charge must be calculated. Be sure and follow the procedures and rules presented in the example problems.

1.   10M sheets that are cut to 5" × 6" from basic size book paper.   _____

2.   15M sheets that are cut to 6" × 10" from basic size bond paper.   _____

3.   25M sheets that are cut to 3" × 5" from basic size bond paper.   _____

4.   2M sheets that are cut to 5" × 7" from basic size bond paper.   _____

5.   3M sheets that are cut to 6½" × 7⅝" from basic size index paper.   _____

6.   12M sheets that are cut to 8" × 10" from basic size book paper.   _____

7.   30M sheets that are cut to 17" × 11" (cut apart only) from basic size bond paper.   _____

8.   45M sheets that are cut to 15½" × 20¾" from basic size cover paper.   _____

9.   35M sheets that are cut to 7" × 5" (cut apart only) from book paper that had previously been cut to 7" × 10".   _____

10.   75M sheets that are cut to 8½" × 11" (back trimmed) and cut from 17½" × 22½" bond paper.   _____

11.   102M sheets that are cut to 5" × 8" from basic size index paper.   _____

12.   450 sheets that are cut to 8½" × 11" from basic size cardboard stock.   _____

13.   800 sheets that are cut to 10" × 12" from basic size index-bristol stock. Note
      that only 100 sheet lifts can be used for this cutting job.   _____

14.   20M sheets that are cut to 8" × 10" from basic size newsprint paper stock.
      Note that the sheets must be back trimmed and that the customer wants
      only 250 sheets cut at one time.   _____

15.   16M sheets that are cut to 5¾" × 10⅜" from basic size ledger paper. Note
      that 1,000 sheets can be cut at one time.   _____

 ## Unit 38   DETERMINING COST OF PAPER STOCK USED

BASIC PRINCIPLES OF DETERMINING COST OF PAPER STOCK USED

Paper is generally sold by either of two systems: *price per hundred weight* (CWT) or *price per 1,000 sheets*. The price per hundred weight (CWT) is the traditional system that has been in place for many years. It is a good system, but it has lost favor in recent years because the price per 1,000 sheet system is much easier. In addition, the price per 1,000 sheet system can be easily calculated without benefit of an extensive formula. The cost of any number of sheets can be easily determined from one sheet up to 100,000 sheets or more, and the only tool needed for a calculation is a basic calculator.

Usually paper prices are included in paper company pricing catalogs according to both pricing systems (Figure 38–1). Upon studying the pricing table, it is possible to determine the cost of paper according to 1,000 sheets in an easy manner. For example, the basic size of 20-pound white bond paper (17" × 22") is priced at $68.96, $45.14, $39.68, and $36.72. These prices are based on 1,000 sheets of paper per the stated size and according to the packaged amount purchased at one time. The more paper purchased at one time, the more economical it is per sheet. This sounds and looks confusing, but it is actually fairly simple once the pricing system is understood.

To determine the cost of one sheet of paper at any listed price, it is only necessary to move the decimal three places to the left. For example, if 1,000 sheets of paper are priced at $68.96, then one sheet will cost $0.06896. This amount per sheet can then be multiplied by the actual number of sheets purchased to find the final cost. The four listed prices of 20-pound white bond paper per basic size are based on how much paper is purchased at one time. The base price is according to one carton, and in the sample pricing sheet, 1,000 sheets of paper purchased in a single carton will cost $45.14, or $0.04514 per sheet. The one confusing aspect is that most of the time more than 1,000 sheets are packaged in a carton. In this example, there are 3,000 sheets in one carton, thus one full carton of the bond paper will cost $0.04514 × 3,000 = $135.42.

Using the pricing chart in Figure 38–1, the following paper prices are provided as examples. Follow the numbers to learn how easy it really is to use a paper pricing catalog.

Example 1:   20-pound white bond paper, 17" × 22"

Plan to purchase 2,000 sheets, thus pricing per broken carton (BKN CTN). (**Note:** The price is higher than the carton price because there is more required handling of the paper.)

$68.96 per M, thus $0.06896 per sheet

$0.06896 × 2,000 = $137.92 total cost

Example 2:   20-pound white bond paper, 24" × 38"

Plan to purchase 13,000 sheets, thus pricing per 4-CTN is possible

$97.21 per M, thus $0.09721 per sheet

$0.09721 × 13,000 = $1,263.73 total cost

Example 3:   20-pound green paper, 22" × 34"

Plan to purchase 16 cartons, thus 1,500 × 16 = 24,000 sheets

$75.20 per M, thus $0.07520 per sheet

$0.07520 × 24,000 = $1,804.80 total cost

Paper pricing tables for other papers besides bond paper are organized in the same manner (Figures 38–2 and 38–3). It is critical to find the correct kind, color, size, and weight of paper in the table so the right price is selected. In addition, it is important to know how many sheets are packaged in one carton and to know how many cartons will be needed so the correct price can be selected.

```
HAMMERMILL BOND
WATERMARKED
BASIS 17 X 22

                       M SHT           SHTS             STOCK   BKN
ITEM      SIZE     GR  WEIGHT  BASIS   CTN    COLOR      LOCATE  CTN    1 CTN   4 CTN 16 CTN
                                                               ──────────── CWT PRICE ────────────

WHITE................ 16 LB.                                    183.50 120.15 105.60  97.70
WHITE................ 20, 24 LB.                                172.40 112.85  99.20  91.80
COLORS............... 16 LB.                                    179.35 117.40 103.20  95.50
COLORS............... 20 LB.                                    176.55 115.60 101.60  94.00
WHITE, EMBOSSED...... 20 LB.                                    187.65 122.85 108.00  99.90
                                                               ──────────── M SHEET PRICE ────────────


20 LB. WHITE-CARTONS
AVG CAL .0046
112006    17 X 22   L    40     20   3000    WHITE      NCH K   68.96  45.14  39.68  36.72
112007    22 X 34   S    80     20   1500    WHITE      N   K  137.92  90.28  79.36  73.44
112009    24 X 38   S    98     20   1500    WHITE      NC     168.95 110.59  97.21  89.96
112010    28 X 34   S   102     20   1500    WHITE      N      175.84 115.10 101.18  93.63


24 LB. WHITE-CARTONS
AVG CAL .0054
112013    22 X 34   S    96     24   1500    WHITE      N      165.50 108.33  95.23  88.12


WRITING PAPERS


HAMMERMILL BOND COLORS
20 LB. COLORS-CARTONS: BLUE, BUFF, CAFE, CHERRY, GOLD,
GRAY, GREEN, IVORY, MELON, ORCHID, PINK, SALMON
AVG CAL .0046

                       M SHT           SHTS             STOCK   BKN
ITEM      SIZE     GR  WEIGHT  BASIS   CTN    COLOR      LOCATE  CTN    1 CTN   4 CTN 16 CTN
                                                               ──────────── M SHEET PRICE ────────────

112041    22 X 34   S    80     20   1500    PINK       N      141.24  92.48  81.28  75.20
112043    22 X 34   S    80     20   1500    SALMON     N H K  141.24  92.48  81.28  75.20
112038    22 X 34   S    80     20   1500    GRAY       N H K  141.24  92.48  81.28  75.20
112026    22 X 34   S    80     20   1500    CANARY     NC     141.24  92.48  81.28  75.20
112017    22 X 34   S    80     20   1500    SKY BLUE   NC     141.24  92.48  81.28  75.20
112035    22 X 34   S    80     20   1500    GREEN      N H    141.24  92.48  81.28  75.20
112042    22 X 34   S    80     20   1500    IVORY      N      141.24  92.48  81.28  75.20
112180    22 X 34   L    80     20   1500    CANARY       H    141.24  92.48  81.28  75.20


20 LB. WHITE RIPPLETONE EMBOSSED
AVG CAL .0041
112011    17 X 22   L    40     20   3000    WHITE      N H    75.06  49.14  43.20  39.96


20 LB. WHITE LINEN EMBOSSED
AVG CAL .0041
112012    22 X 34   S    80     20   1500    WHITE      NC     150.12  98.28  86.40  79.92
```

Figure 38–1 A typical pricing schedule for one of the many bond papers. (Credit: Hammermill Papers.)

UNCOATED BOOK

CARNIVAL BOOK

	PRICE PER CWT.			
	BKN. CTN.	*1* CTN.	*4* CTN.	*16* CTN.
50 lb. White, Smooth & Vellum	156.70	89.55	77.60	71.65
60 lb. & Heavier White, Smooth & Vellum	151.90	86.80	75.25	69.45
White, Lynnfield & Handmade	164.70	94.10	81.55	75.25
White, Linen	171.00	97.70	84.70	78.20
Standard Colors, Vellum	161.45	92.25	79.95	73.80
Orange, Vellum	174.20	99.55	86.25	79.65
Lime, Vellum	206.00	117.70	102.00	94.20

Basis 25 × 38	Size	M Wt.	Ctn. Pack	PRICE PER 1000 SHEETS			
				BKN. CTN.	*1* CTN.	*4* CTN.	*16* CTN.
WHITE, SMOOTH							
50	17½ × 22½	41	3600	64.25	36.72	31.82	29.38
	23 × 35	85	1800	133.20	76.12	65.96	60.90
60	17½ × 22½	50	3200	75.95	43.40	37.63	34.73
	19 × 25	60	2400	91.14	52.08	45.15	41.67
	23 × 29	84	1800	127.60	72.91	63.21	58.34
	23 × 35	102	1500	154.94	88.54	76.76	70.84
	25 × 38	120	1200	182.28	104.16	90.30	83.34
	35 × 45	198	800	300.76	171.86	149.00	137.51
70	17½ × 22½	58	2400	88.10	50.34	43.65	40.28
	23 × 29	98	1600	148.86	85.06	73.75	68.06
	23 × 35	119	1200	180.76	103.29	89.55	82.65
	25 × 38	140	1000	212.66	121.52	105.35	97.23
	35 × 45	232	600	352.41	201.38	174.58	161.12
80	23 × 35	136	1100	206.58	118.05	102.34	94.45
	25 × 38	160	1000	243.04	138.88	120.40	111.12
100	23 × 29	140	1000	212.66	121.52	105.35	97.23
	23 × 35	169	900	256.71	146.69	127.17	117.37
	25 × 38	200	800	303.80	173.60	150.50	138.90
	35 × 45	332	500	504.31	288.18	249.83	230.57
COLORS, VELLUM							
50	17½ × 22½	41	3600	64.25	36.72	31.82	29.38
	25 × 38	100	1600	156.70	89.55	77.60	71.65
	35 × 45	166	900	260.12	148.65	128.82	118.94
60	17½ × 22½	50	3200	75.95	43.40	37.63	34.73
	23 × 35	102	1500	154.94	88.54	76.76	70.84
	25 × 38	120	1200	182.28	104.16	90.30	83.34
	35 × 45	198	800	300.76	171.86	149.00	137.51
70	17½ × 22½	58	2400	88.10	50.34	43.65	40.28
	23 × 35	119	1200	180.76	103.29	89.55	82.65
	25 × 38	140	1000	212.66	121.52	105.35	97.23
	35 × 45	232	600	352.41	201.38	174.58	161.12

Figure 38–2 A typical pricing schedule for one of the many book papers.

UNCOATED BOOK

CARNIVAL COVER ANTIQUE

Basis 20 × 26	Size	M Wt.	Ctn. Pack	PRICE PER 1000 SHEETS			
				BKN. CTN.	1 CTN.	4 CTN.	16 CTN.
WHITE							
50	20 × 26	100	1500	156.70	89.55	77.60	71.65
	23 × 35	155	1000	242.89	138.80	120.28	111.06
65	20 × 26	130	1000	201.71	116.42	100.88	93.15
	23 × 35	201	750	314.97	180.00	155.98	144.02
	26 × 40	260	500	407.42	232.83	201.76	186.29
STANDARD COLORS Blue, Goldenrod, Light Green, India, Ivory, Putty, Sand, Sky, Yellow							
65	20 × 26	130	1000	216.13	123.50	106.99	98.80
	23 × 35	201	750	334.16	190.95	165.42	152.76
	26 × 40	260	500	432.25	247.00	213.98	197.60
DEEP COLORS Orange, Medium Blue, Brown							
65	20 × 26	130	1000	232.64	132.93	115.18	106.34
	23 × 35	201	750	359.69	205.52	178.09	164.42
	26 × 40	260	500	465.27	265.85	230.36	212.68
GREEN, LIME, RED							
65	20 × 26	130	1000	274.04	156.59	135.72	125.26
	23 × 35	201	750	423.71	242.10	209.84	193.66
	26 × 40	260	500	548.08	313.17	271.44	250.51

CAROLINA COATED 1 SIDE COVER

Basis Caliper	Size	M Wt.	Ctn. Pack	PRICE PER 1000 SHEETS			
				BKN. CTN.	1 CTN.	4 CTN.	16 CTN.
WHITE							
8 Pt.	23 × 35	212	700	296.80	169.60	147.02	135.68
10 Pt.	20 × 26	160	900	203.60	116.32	100.80	93.12
	23 × 35	248	600	315.58	180.30	156.24	144.34
	25 × 38	292	500	371.57	212.28	183.96	169.94
	26 × 40	320	500	407.20	232.64	201.60	186.24

Figure 38–3 A typical pricing schedule for one of the many cover papers.

PRACTICAL PROBLEMS

Note: For the specifications given in problems 1–15, the cost of the paper must be calculated. The pricing tables in Figures 38–1, 38–2, and 38–3 are to be used for determining the paper costs for these problems. Be sure to follow the procedures and rules presented in the example problems.

1.   4,500 sheets of 17" × 22", 20-lb. white bond paper.   _____

2.   50,000 sheets of 17" × 22", 20-lb. white bond paper.   _____

3.   10,000 sheets of 28" × 34", 20-lb. white bond paper.   _____

4.   7,000 sheets of 24" × 38", 20-lb. white bond paper.   _____

5.   1,350 sheets of 22" × 34", 20-lb. white bond paper.   _____

6.   750 sheets of 22" × 34", 20-lb. gold bond paper.   _____

7.   23,500 sheets of 22" × 34", 20-lb. salmon bond paper.   _____

8.   5,000 sheets of 17" × 22" and 5,000 sheets of 24" × 38", 20-lb. white bond paper.   _____

9.   6,500 sheets of 25" × 38", 50-lb. white uncoated book paper.   _____

10.   30,000 sheets of 35" × 45", 60-lb. white uncoated book paper.   _____

11.   75,000 sheets of 20" × 26", 65-lb. green uncoated cover paper.   _____

12.   135,000 sheets of 26" × 40", 65-lb. blue uncoated cover paper.   _____

13.   4,750 sheets of 25" × 38", 60-lb. white uncoated book paper and 9,500 sheets of 17½" × 22½", 50-lb. white uncoated book paper.   _____

14.   13,350 sheets of 23" × 35", 60-lb. white uncoated book paper and 15,825 sheets of 35" × 45", 60-lb. white uncoated book paper.   _____

15.   50,000 sheets of 23" × 35", 50-lb. white uncoated book paper and 95,000 sheets of 20" × 26", 10-pt. white coated-one-side cover paper.   _____

Appendix

PRINTER'S UNIT
CUSTOMARY & METRIC
EQUIVALENTS

PRINTER'S		CUSTOMARY		METRIC
		Inches		
Picas	Points	Fraction	Decimal	Millimeters
	1	$1/64$	.014	.35
	2	$1/32$	.028	.70
	3	$3/64$	.042	1.05
	4	$7/128$	.055	1.40
	5	$1/16$	.069	1.75
	6	$5/64$	.083	2.10
	7	$3/32$	.097	2.45
	8	$7/64$	.111	2.80
	9	$1/8$	.125	3.15
	10	$9/64$	.138	3.50
1	12	$21/128$	.166	4.20
	14	$25/128$	.194	4.90
	18	$1/4$	.249	6.30
2	24	$21/64$	.332	8.40
	30	$53/128$	.414	10.50
3	36	$1/2$	.498	12.60
	42	$37/64$	.581	14.70
4	48	$85/128$	.664	16.80
5	60	$53/64$	.828	21.00
6	72	1	.996	25.20

CONVERSION OF PICAS TO INCHES

Picas	Inches	Picas	Inches	Picas	Inches	Picas	Inches
1	.166	26	4.316	51	8.466	76	12.616
2	.332	27	4.482	52	8.632	77	12.782
3	.498	28	4.648	53	8.798	78	12.948
4	.664	29	4.814	54	8.964	79	13.114
5	.830	30	4.980	55	9.130	80	13.280
6	.996	31	5.146	56	9.296	81	13.446
7	1.162	32	5.312	57	9.462	82	13.612
8	1.328	33	5.478	58	9.628	83	13.778
9	1.494	34	5.644	59	9.794	84	13.944
10	1.660	35	5.810	60	9.960	85	14.110
11	1.826	36	5.976	61	10.126	86	14.276
12	1.992	37	6.142	62	10.292	87	14.442
13	2.158	38	6.308	63	10.458	88	14.608
14	2.324	39	6.474	64	10.624	89	14.774
15	2.490	40	6.640	65	10.790	90	14.940
16	2.656	41	6.806	66	10.956	91	15.106
17	2.822	42	6.972	67	11.122	92	15.272
18	2.988	43	7.138	68	11.288	93	15.438
19	3.154	44	7.304	69	11.454	94	15.604
20	3.320	45	7.470	70	11.620	95	15.770
21	3.486	46	7.636	71	11.786	96	15.936
22	3.652	47	7.802	72	11.952	97	16.102
23	3.818	48	7.968	73	12.118	98	16.268
24	3.984	49	8.134	74	12.284	99	16.434
25	4.150	50	8.300	75	12.450	100	16.600

DECIMAL CHART

inches fractional	decimal	inches fractional	decimal
1/64	.0156	33/64	.5156
1/32	.0312	17/32	.5312
		35/64	.5469
3/64	.0469	9/16	.5625
1/16	.0625		
5/64	.0781	37/64	.5781
3/32	.0938	19/32	.5938
7/64	.1094	39/64	.6094
		5/8	.6250
1/8	.1250		
9/64	.1406	41/64	.6406
5/32	.1562	21/32	.6562
11/64	.1719	43/64	.6719
3/16	.1875	11/16	.6875
		45/64	.7031
13/64	.2031		
7/32	.2188	23/32	.7188
15/64	.2344	47/64	.7344
1/4	.2500	3/4	.7500
17/64	.2656	49/64	.7656
		25/32	.7812
9/32	.2812		
19/64	.2969	51/64	.7969
5/16	.3125	13/16	.8125
21/64	.3281	53/64	.8281
11/32	.3438	27/32	.8438
		55/64	.8594
23/64	.3594		
3/8	.3750	7/8	.8750
25/64	.3906	57/64	.8906
13/32	.4062	29/32	.9062
27/64	.4219	59/64	.9219
		15/16	.9375
7/16	.4375		
29/64	.4531	61/64	.9531
15/32	.4688	31/32	.9688
31/64	.4844	63/64	.9844
1/2	.5000	1	1.0000

m/m	inches	m/m	inches	m/m	inches	m/m	inches	m/m	inches	m/m	inches
1	0.0394	51	2.0079	101	3.9764	151	5.9449	201	7.9134	251	9.8819
2	0.0787	52	2.0472	102	4.0157	152	5.9843	202	7.9527	252	9.9212
3	0.1181	53	2.0866	103	4.0551	153	6.0236	203	7.9921	253	9.9606
4	0.1575	54	2.1260	104	4.0945	154	6.0630	204	8.0315	254	10.0000
5	0.1969	55	2.1654	105	4.1339	155	6.1024	205	8.0709	255	10.0394
6	0.2362	56	2.2047	106	4.1732	156	6.1417	206	8.1102	256	10.0787
7	0.2756	57	2.2441	107	4.2126	157	6.1811	207	8.1496	257	10.1181
8	0.3150	58	2.2835	108	4.2520	158	6.2205	208	8.1890	258	10.1575
9	0.3543	59	2.3228	109	4.2913	159	6.2598	209	8.2283	259	10.1968
10	0.3937	60	2.3622	110	4.3307	160	6.2992	210	8.2677	260	10.2362
11	0.4331	61	2.4016	111	4.3701	161	6.3386	211	8.3071	261	10.2756
12	0.4724	62	2.4409	112	4.4094	162	6.3780	212	8.3464	262	10.3149
13	0.5118	63	2.4803	113	4.4488	163	6.4173	213	8.3858	263	10.3543
14	0.5512	64	2.5197	114	4.4882	164	6.4567	214	8.4252	264	10.3937
15	0.5906	65	2.5591	115	4.5276	165	6.4961	215	8.4646	265	10.4331
16	0.6299	66	2.5984	116	4.5669	166	6.5354	216	8.5039	266	10.4724
17	0.6693	67	2.6378	117	4.6063	167	6.5748	217	8.5433	267	10.5118
18	0.7087	68	2.6772	118	4.6457	168	6.6142	218	8.5827	268	10.5512
19	0.7480	69	2.7165	119	4.6850	169	6.6535	219	8.6220	269	10.5905
20	0.7874	70	2.7559	120	4.7244	170	6.6929	220	8.6614	270	10.6299
21	0.8268	71	2.7953	121	4.7638	171	6.7323	221	8.7008	271	10.6693
22	0.8661	72	2.8346	122	4.8031	172	6.7717	222	8.7401	272	10.7086
23	0.9055	73	2.8740	123	4.8425	173	6.8110	223	8.7795	273	10.7480
24	0.9449	74	2.9134	124	4.8819	174	6.8504	224	8.8189	274	10.7874
25	0.9843	75	2.9528	125	4.9213	175	6.8898	225	8.8583	275	10.8268
26	1.0236	76	2.9921	126	4.9606	176	6.9291	226	8.8976	276	10.8661
27	1.0630	77	3.0315	127	5.0000	177	6.9685	227	8.9370	277	10.9055
28	1.1024	78	3.0709	128	5.0394	178	7.0079	228	8.9764	278	10.9449
29	1.1417	79	3.1102	129	5.0787	179	7.0472	229	9.0157	279	10.9842
30	1.1811	80	3.1496	130	5.1181	180	7.0866	230	9.0551	280	11.0236
31	1.2205	81	3.1890	131	5.1575	181	7.1260	231	9.0945	281	11.0630
32	1.2598	82	3.2283	132	5.1969	182	7.1654	232	9.1338	282	11.1023
33	1.2992	83	3.2677	133	5.2362	183	7.2047	233	9.1732	283	11.1417
34	1.3386	84	3.3071	134	5.2756	184	7.2441	234	9.2126	284	11.1811
35	1.3780	85	3.3465	135	5.3150	185	7.2835	235	9.2520	285	11.2205
36	1.4173	86	3.3858	136	5.3543	186	7.3228	236	9.2913	286	11.2598
37	1.4567	87	3.4252	137	5.3937	187	7.3622	237	9.3307	287	11.2992
38	1.4961	88	3.4646	138	5.4331	188	7.4016	238	9.3701	288	11.3386
39	1.5354	89	3.5039	139	5.4724	189	7.4409	239	9.4094	289	11.3779
40	1.5748	90	3.5433	140	5.5118	190	7.4803	240	9.4488	290	11.4173
41	1.6142	91	3.5827	141	5.5512	191	7.5197	241	9.4882	291	11.4567
42	1.6535	92	3.6220	142	5.5906	192	7.5591	242	9.5275	292	11.4960
43	1.6929	93	3.6614	143	5.6299	193	7.5984	243	9.5669	293	11.5354
44	1.7323	94	3.7008	144	5.6693	194	7.6378	244	9.6063	294	11.5748
45	1.7717	95	3.7402	145	5.7087	195	7.6772	245	9.6457	295	11.6142
46	1.8110	96	3.7795	146	5.7480	196	7.7165	246	9.6850	296	11.6535
47	1.8504	97	3.8189	147	5.7874	197	7.7559	247	9.7244	297	11.6929
48	1.8898	98	3.8583	148	5.8268	198	7.7953	248	9.7638	298	11.7323
49	1.9291	99	3.8976	149	5.8661	199	7.8346	249	9.8031	299	11.7716
50	1.9685	100	3.9370	150	5.9055	200	7.8740	250	9.8425	300	11.8110

METRIC UNITS

Measures of Length		
10 millimeters	=	1 centimeter, cm
10 centimeters	=	1 decimeter, dm
10 decimeters	=	1 meter, m
10 meters	=	1 decameter, dam
10 decameters	=	1 hectometer, hm
10 hectometers	=	1 kilometer, km
10 kilometers	=	1 myriameter, mym
Measures of Weight		
10 milligrams, mg	=	1 centigram, cg
10 centigrams	=	1 decigram, dg
10 decigrams	=	1 gram, g
10 grams	=	1 decagram, dag
10 decagrams	=	1 hectogram, hg
10 hectograms	=	1 kilogram, kg
10 kilograms	=	1 myriagram, myg
10 myriagrams	=	1 quintal, q
10 quintals	=	1 millier or ton, MT or t
Measures of Liquid Volume		
10 milliliters, ml	=	1 centiliter, cl
10 centiliters	=	1 deciliter, dl
10 deciliters	=	1 liter, l
10 liters	=	1 decaliter, dal
10 decaliters	=	1 hectoliter, hl
10 hectoliters	=	1 kiloliter, kl

APPROXIMATE CONVERSIONS

When you know	Multiply by	To find	Symbol
		(10 millimeters equals 1 centimeter)	
inches	2.5	centimeters	cm
feet	30.0	centimeters	cm
yards	0.9	meters	m
miles	1.6	kilometers	km
square inches	6.5	square centimeters	cm^2
square feet	0.09	square meters	m^2
square yards	0.8	square meters	m^2
square miles	2.6	square kilometers	km^2
acres	0.4	hectares	ha
ounces	28.0	grams	g
pounds	0.45	kilograms	kg
short tons (2000 lb)	0.9	tonnes	t
teaspoons	5.0	milliliters	ml
tablespoons	15.0	milliliters	ml
fluid ounces	30.0	milliliters	ml
cups	0.24	liters	l
pints	0.47	liters	l
quarts	0.95	liters	l
gallons	3.8	liters	l
cubic feet	0.03	cubic meters	m^3
cubic yards	0.76	cubic meters	m^3
Fahrenheit Temperature	5/9 (after subtracting 32)	Celsius Temperature	°C

Glossary

Note: The number in parentheses following the glossary term indicates the section introduction or unit in which the term is first used in this textbook.

+ symbol — (sec. 1) A mathematical symbol indicating that two or more number groups should be added together.

- symbol — (sec. 1) A mathematical symbol indicating that one number group should be subtracted from another number group.

x symbol — (sec. 1) A mathematical symbol indicating that two or more number groups should be multiplied by each other.

÷ symbol — (sec. 1) A mathematical symbol indicating that one number group should be divided into another number group.

= symbol — (sec. 1) A mathematical symbol indicating that one or more number groups will equal a numerical amount based on one or more previous actions.

% symbol — (sec. 4) A mathematical symbol indicating that a number represents a percent of 100%.

Addition — (unit 1) The mathematical function of combining two or more numbers into one larger number.

Aesthetics — (sec. 7) The area of philosophical theory relating to whether something is or is not attractive and/or beautiful.

Annum — (unit 19) A term often used by mathematicians and accountants to indicate a one-year period of twelve months.

Anvil — (unit 25) An important part to the standard micrometer: the base upon which the spindle is aligned. Material being measured is placed between the anvil and spindle.

Arabic number — (unit 6) The name given to the numbers 0 through 9. See *whole number*.

Area — (unit 21) A term relating to the measurement of a given set of dimensions: for ex-

ample, a rectangle that is 2 feet by 3 feet has an area of 6 square feet.

Back trimming — (sec. 10) The procedure of trimming all four sides of stock sheets when creating press sheets.

Basis weight — (unit 32) The weight of one ream of paper per its basic size.

Bleed — (unit 7) An image that is positioned so it will run off the page on one, two, three, or four sides when the sheet is trimmed.

Caliper — (unit 32) The thickness of paper and paperboard as measured in thousandths of an inch. See *Points*.

Centimeter — (unit 23) A unit of measure within the metric system of measurement; one of 100 parts in a meter.

Chart — (sec. 6) A source of information, usually, designed in tabular format. A chart can contain written information, numerical information, or a combination of both.

Commodity — (sec. 9) A product of economic value, such as printing ink or paper.

Common fraction — (sec. 2) A mathematical character containing two sets of numbers separated by a horizontal or slanted line that represents a part of a whole number: for example, ½, ¼, ⅓, and so on.

Convert — (unit 16) To alter numbers from one condition to another; to change a number from a decimal fraction to a common fraction and vice versa.

Customary measurement system — (sec. 5) A standard measurement system of weights and measures in which the inch, foot, yard, quart, and gallon are used as measures of consistency.

Cut apart only — (sec. 10) Making one cut in a stock sheet to create two ready-to-use press sheets.

Cyan — (unit 7) One of the three process ink colors created from the primary light colors of blue and green.

Decimal — (sec. 3) A dot or period used to identify the separation of whole numbers from partial numbers: for example, in 1.5, 1 is a whole number, and 5 is one-half of a whole number.

Decimal fraction — (sec. 3) A fraction of the whole after which the denominator is the power of 10 when a dot is placed to the left of a whole number: for example, .2 = $^2/_{10}$, .05 = $^5/_{100}$, .025 = $^{25}/_{1000}$.

Decimeter — (unit 23) A unit of measure within the metric system of measurement; one of 10 parts in a meter.

Denominator — (sec. 2) The number of a common fraction that is below the fraction bar or the line separating the numerator and the denominator.

Difference — (unit 2) The amount that numbers differ in quantity or measure, such as when one number is subtracted from another number.

Digit — (sec. 1) Any one of the numbers 0 through 9 that, when used singularly or combined in any combination, forms numerical amounts in the mathematical system.

Dividend — (unit 4) The number that is to be divided by the divisor; the base number that will be divided by a whole number, common fraction, or decimal fraction.

Division — (unit 4) The mathematical procedure of dividing or reducing a given number into two or more parts: for example, 10 ÷ 5 = 2.

Divisor — (unit 10) The number or quantity by which the dividend is divided to produce the quotient. Divisor numbers are used when dividing whole numbers, fraction numbers, and decimal fractions.

Dot — (sec. 3) A small spot or point made with a sharpened instrument such as a pencil or the correct stroke of a computer key. It is used in mathematics to show positions of numbers.

Enumeration — (unit 6) The procedure of placing one number after another: for example, 1, 2, 3, and so on.

Equals — (unit 18) In mathematics, *is* and *equals* carry the same meaning as the = symbol: for example, .5 is equal to ½, .25 is equal to ¼, and so on.

First quarter — (unit 12) A period of time representing the first three months (January, February, and March) or thirteen weeks of any year. It is often used by accountants and business executives to compare business sales and/or assets to the same time period of the preceding year.

Fraction — (sec. 2) A numerical representation of a number less than the whole stated in either a common fraction form or a decimal fraction form.

Fraction bar — (sec. 2) The line, either in a horizontal or diagonal position, that separates the numerator from the denominator in a common fraction.

Gang run — (unit 34) Imaging two or more finished sheets on one press sheet. Once imaged, the press sheet is cut apart to create two or more correctly sized finished sheets.

Ganged — (unit 9) Two or more finished sheets that are placed on one press sheet. See *Gang run*.

Grain of paper — (unit 32) The direction of the wood and cotton fibers that constitute paper and paperboard. Paper grain is either the width or length of the paper. Paper will fold easier with the grain.

Gram — (unit 23) A standard unit of weight measurement within the metric measurement system.

Graph — (sec. 6) A diagram or image that has been created to compare two or more bits of information using lines, points, and bars.

Gripper margin — (unit 7) The leading edge of a piece of substrate (paper) when it is being fed or run through a printing press or photocopy machine.

Hardware — (sec. 7) A term that most frequently refers to computer equipment, such as the CPU (central processing unit), monitor, and keyboard; it is also used when referring to computer accessories such as scanners and printers.

Hundredths position — (sec. 3) The second number position following the decimal when listed in a decimal fraction: for example, in 5.284, 8 is in the hundredths position.

Hundred-thousandths position — (sec. 3) The fifth number position following the decimal when listed in a decimal fraction: for example, in 5.28463, 3 is in the one-hundredth-thousandths position.

Illustration — (sec. 6) A drawing of an image that helps communicate an idea or thought that might take hundreds and even thousands of words to describe.

Improper fraction — (unit 7) A condition of the numerator being larger than the denominator caused when adding or multiplying common fractions.

Ink coverage chart — (unit 35) A chart containing information relative to different categories of copy and how much area selected copy will cover from 15% to 100%.

Ink mileage factor — (unit 36) A chart containing information relative to the square inches of area that can be covered by one pound of printing ink on various printing papers.

Interest — (unit 19) A monetary charge for borrowing money: for example, 5% interest for borrowing $200 for 12 months equals $10.

Kilogram — (unit 23) A standard unit of weight measurement within the metric measurement system; 1,000 grams are equal to one kilogram.

Leading — (unit 22) The amount of physical space placed between lines of type. Leading is usually measured in points, such as one point of leading should be used in a given typeform.

Lift — (sec. 10) See *Lift of paper*.

Lift of paper — (unit 32) The amount of paper that can be conveniently and accurately handled by hand or machine when performing selected operations such as cutting and trimming.

Line gauge — (unit 22) A measuring device or tool equal to the foot ruler and the yardstick; common units of measure printed or stamped on a line gauge include points, half-picas, picas, and inches.

Line graph — (unit 26) An illustration created from data of two or more entities showing the comparison between two or among several bits of information; a communication form in which lines are used to connect various high to low points.

Linear — (unit 21) A term relating to a single dimension of measurement such as length; a straight line or a single line of thinking.

Liquid — (unit 21) A term relating to volume measurement, such as a quart of milk: for example, two pints equal one quart, and four quarts equal one gallon.

Liter — (unit 23) The standard unit of liquid measurement in the metric measurement system; slightly larger in amount than the standard quart.

Long division — (unit 4) Dividing one number by another number by hand as compared to using a calculator to make the mathematical calculation.

Magenta — (unit 7) One of the three process ink colors created from the primary light colors of red and blue.

Makeready — (unit 34) The preparation required for making a machine ready for use; adjusting a printing press for the paper size and thickness including registration of the image.

Measurement — (sec. 5) The act or procedure of measuring something to determine its size according to an agreed-upon system of sizes and amounts.

Meter — (unit 23) The basic metric measurement unit of length.

Metric measurement system — (sec. 5) A standard decimal measuring system that is based on the meter and the kilogram.

Micrometer — (sec. 5) A precision measuring device that is usually designed for providing measurements in thousandths of an inch.

Milliliter — (unit 23) A unit of liquid measure within the metric system of measurement; one of 1,000 parts of a liter.

Millimeter — (unit 23) A unit of measure within the metric system of measurement; one of 1,000 parts in a meter.

Mixed number — (unit 7) A numerical amount containing both a whole number and a fraction number: for example, 2½ is a mixed number.

Monetary discount — (unit 20) An amount of money, most often represented by a specific percentage, that will be discounted from an original price: for example, 10% discount if purchased on Monday or Tuesday.

Multiplication — (unit 3) A mathematical procedure that increases one number by the number of times according to the size of the second number: for example, $10 \times 3 = 30$.

Number — (sec. 1) A symbol of communication representing an amount of something: for example, 10 is a number signifying that there are ten single items totaling the number 10.

Numerator — (sec. 2) The number of a common fraction that is above the fraction bar or the line separating the numerator and the denominator.

Page creation software — (unit 28) Special computer software that has been written for the specific purpose of creating pages of copy that include illustrations, photographs, and type.

Part method — (unit 30) A term that refers to establishing visual balance when positioning an illustration, picture, or a typeform on a page or in a frame, using the margin ratios of 3 parts at the top, 4 parts on both sides, and 5 parts at the bottom.

Percent — (sec. 4) A procedure to determine the portion of the whole that is 100%; for example, 90% (of 100%) of the project was completed on time.

Percentage — (sec. 4) A figure representing part of a whole: for example, 90% is 9/10 of the whole.

Photograph — (sec. 6) A picture or likeness of an object, person, or landscape that is created through the process of photography.

Pica — (unit 7) A unit of measure within the point measurement system. One pica is equal to 12 points, and there are approximately 6 picas in one inch.

Pie chart — (unit 26) An illustration made in the shape of a 360° circle and divided into pieces, somewhat like the slices of a pie. A pie chart is used to show a comparison of the sizes of the various parts to the whole.

Point — (sec. 5) A very small measurement unit; 1/72 of a customary inch measurement.

Point measurement system — (sec. 5) A stan- dard measurement system used primarily to measure typefaces and distances within typeforms and pages of type characters. There are approximately 72 points in a standard inch measure.

Points — (unit 32) A term used with reference to measuring thick papers and paperboards in thousandths of an inch; a point of paper thickness represents 1/1000 inch. See *Caliper.*

Press sheets — (sec. 8) Sheets of paper stock or other substrate that have been cut from stock sheets to the correct size for running through a printing press or photocopier.

Price per hundred weight — (unit 38) The traditional method of pricing and selling printing and other papers that are sold in roll form; selling various papers according to 100 pounds of weight.

Price per 1,000 sheets — (unit 38) The accepted method of selling cut sheet printing and other papers; selling various papers according to 1,000 sheets, which can be reduced to price per sheet by moving the decimal three places to the left.

Principal — (unit 19) A capital sum of money that represents the total amount due or that is borrowed.

Product — (unit 3) The result or number that is obtained when two numbers are multiplied: for example, in 10 × 3 = 30, 30 is the product.

Proportion — (sec. 5) The relationship of one part to another part or to the whole with respect to size, quantity, or degree.

Proportional dial — (unit 27) A device used to determine the enlargement and reduction sizes of copy. See *Proportional wheel.*

Proportional wheel — (unit 27) A device or tool containing two wheel or dials with measurement markings that are fastened together in the middle so the two dials turn either way. It is used to determine the enlargement and reduction sizes of an illustration or photograph.

Pulpwood — (sec. 8) Trees that have grown to the correct size for cutting and for being used to make paper and paperboard.

Quotient — (unit 4) The result or number that is acquired when one number is divided by another number: for example, in 10 ÷ 2 = 5, 5 is the quotient.

Ratio — (sec. 5) A relationship in quantity, amount, or size of one thing to another thing.

Readability — (sec. 7) A measure of how well a typeface classification can be read by the human eye.

Ream — (unit 1) An amount of paper; 500 sheets of paper of any size, kind, and color.

Reciprocal — (unit 10) The result of reversing the numerator and denominator of a common fraction: for example, the reciprocal of ⅝ is ⅝.

Remainder — (unit 2) The amount or number representing what is left after the subtraction procedure has been completed.

Re-size — (unit 27) To change the size of an illustration, photograph, or type based on the original size; to make an image larger or smaller as needed.

Roman numeral — (unit 6) A numeral or number that is based on the mathematical system developed by the ancient Roman people: for example, I = 1, V = 5, X = 10, and so on.

Rounding to decimal places — (sec. 3) The procedure of increasing a preceding number or leaving it the same at a given decimal position: for example, rounding 2.846 at the second place or hundredths position would make the decimal fraction 2.85.

Set 8 on 10 — (unit 29) A typographer's markup terminology to inform the typesetter to set the type in 8 points with 10 points between the baselines. This designation is a form of communication to the typesetter that there should be two points of leading between the lines.

Set-width — (unit 22) The width of a single type character. The capital letter A, for example, is a given number of points wide depending on the type size being used.

Signature — (unit 2) A folded sheet of paper that has been imaged and contains four or more individual pages. Typical signature page counts include 4, 8, 12, 16, 32, and 64.

Sizing — (sec. 6) Changing the physical size by decreasing or increasing the original size of an illustration, photograph, or letters to fit a given space.

Software — (sec. 7) A term that is frequently used when referring to computer programs that have been written to make computer hardware actually function; very intricate and detailed instructions for the computer to follow based on the input from the keyboard, mouse, drives, and touch screens.

Spindle — (unit 25) An important part of the standard micrometer; the shaft that is moved back and forth to press upon the material being measured.

Spoilage — (unit 34) Substrate (paper and other materials that are imaged) that is spoiled during a specific operation; paper that has been misprinted or damaged during a press run.

Square centimeter — (unit 23) An amount of space that measures one centimeter on each of the four sides.

Stock sheets — (sec. 8) The larger sheets of paper or other substrate from which press sheets are cut and run through a printing press or photocopier.

Substrate — (sec. 8) A substance or material; a term used to identify the several materials, such as paper, wood, metal, glass, plastic, and so on, that are imaged in the graphic communications industry.

Subtraction — (unit 2) The mathematical procedure of removing a given amount or number from another number.

Template form — (unit 26) A basic layout written into selected computer software. Information in the form of type and numbers can be entered into template forms and printed out as one unit.

Ten-thousandths position — (sec. 3) The fourth number position following the decimal when listed in a decimal fraction: for example, in 5.28463, 6 is in the ten-thousandths position.

Tenths position — (sec. 3) The first number position following the decimal when listed in a decimal fraction: for example, in 5.284, 2 is in the tenths position.

Thousandths of an inch — (unit 32) A measurement method in which the common inch has been divided into 1,000 parts. Critical measurements are made in thousandths of an inch. See *Micrometer*.

Thousandths position — (sec. 3) The third number position following the decimal when listed in a decimal fraction: for example, in 5.2846, 4 is in the thousandths position.

Times — (Unit 18) A mathematical term meaning to multiply one number by another. The word *by* when used in a mathematical situation also means to multiply one number by another.

Uncommon fractions — (unit 7) Two or more fraction numbers that do not have the same denominator. To perform mathematical functions, the uncommon fractions must be made into common fractions having the same denominator.

Visual balance — (unit 30) The result of positioning images on a page or in a frame that makes the viewed image look vertically bal-anced; placing more margin space at the bottom of an image than at the top.

Weight — (unit 21) A term relating to the measurement of how much something weighs; the measurement of how heavy, based on its specific gravity, is an object or living being.

Whole number — (sec. 1) Numerical numbers or digits 0 through 9, which can be combined in different ways to form numbers from zero to several million and beyond.

ANSWERS TO ODD-NUMBERED PROBLEMS

SECTION 1   WHOLE NUMBERS

UNIT 1   ADDITION OF WHOLE NUMBERS

1. 65 hours
3. 2,422 reams of paper
5. 12,186 sheets of paper
7. 125,011 tags
9. 2,680 column lines of type
11. 794,142 magazines
13. 58,104 letterheads
15. 26,619 books

UNIT 2   SUBTRACTION OF WHOLE NUMBERS

1. $1,350.00 of profit
3. 14,500 sheets of bond paper
5. 626 column lines of type
7. 7,830 signatures
9. $218.00 lost bid
11. 32 more jobs
13. 165 column lines of type
15. 22 hours

UNIT 3   MULTIPLICATION OF WHOLE NUMBERS

1. $84.00 for order
3. $8,650.00 for paper
5. 30,000 letterheads
7. 38,016 column lines of type
9. 381,480 cards
11. $192.00 labor cost
13. 168,000 impressions
15. $1,642.65 cost of labor

UNIT 4   DIVISION OF WHOLE NUMBERS

1. 1,875 stock sheets
3. 917 reams of paper
5. $598.00 of benefit
7. 4,275 flyers per hour
9. $16.42 average selling price
11. 9 inches each piece
13. 4 ounces per pamphlet
15. 43 reams of paper

UNIT 5   COMBINED OPERATIONS WITH WHOLE NUMBERS

1. $179.00 checkbook balance
3. 46 fuel fillings for month
5. $1,060 total cost of gas
7. 6 computers to one computer
9. 848 average run length
11. $154 for lunches/refreshments
13. 220 copies per hour
15. 13 companies per county